EINSTEIN'S UNFINISHED SYMPHONY

Also by Marcia Bartusiak
Thursday's Universe
Through a Universe Darkly
Archives of the Universe
The Day We Found the Universe
Black Hole

MARCIA BARTUSIAK

Einstein's Unfinished Symphony

THE STORY OF A GAMBLE, TWO BLACK HOLES, AND A NEW AGE OF ASTRONOMY

Yale

UNIVERSITY PRESS

NEW HAVEN & LONDON

Published with assistance from the Louis Stern Memorial Fund.

First Yale University Press edition 2017.
An earlier edition of this book was published in 2000 by Joseph Henry
Press as *Einstein's Unfinished Symphony: Listening to the Sounds of Space-Time*.

Yale University Press books may be purchased in quantity for educational,
business, or promotional use. For information, please e-mail sales.press@
yale.edu (US office) or sales@yaleup.co.uk (UK office).

Unless otherwise noted, interior illustrations are by Rob Wood and
Matthew Frey, Wood Ronsaville Harlin, Inc., Annapolis, Md.

Set in Scala type by IDS Infotech, Ltd.
Printed in the United States of America.

Library of Congress Control Number: 2016957559
ISBN 978-0-300-22339-2

A catalogue record for this book is available from the British Library.

This paper meets the requirements of ANSI/NISO Z39.48-1992
(Permanence of Paper).

10 9 8 7 6 5 4 3 2 1

To the LIGO staff, technicians, and scientists,
for their unflagging spirit through decades of groundwork

CONTENTS

ACKNOWLEDGMENTS TO THE 2017 EDITION

AFTER I PUBLISHED THE FIRST EDITION OF this book, it took fifteen long years for gravitational-wave astronomers to detect their first signal. Everyone said that it would likely occur once the Laser Interferometer Gravitational-Wave Observatory had installed its second generation of instrumentation. The first generation that I had earlier reported on served as a test-bed for the new technologies. What I didn't count on was the signal being captured soon after the advanced, more sensitive equipment was turned on. It caught me off guard, but what a pleasant surprise it turned out to be. It allowed me to add a grand finale (or at least a "chirp") to my symphony.

Much has happened over the years between these editions. Some of the major players in gravitational physics whom I originally wrote about or interviewed passed away, including John Wheeler, Vladimir Braginsky, Joseph Weber, and Roland Schilling. Felix Pirani, whose theoretical work in the 1950s was influential in convincing the physics community that gravity waves were measurable, died just weeks before LIGO scientists announced their first detection. Others have retired, while many of the young graduate students and post-docs, who carried out myriad tasks during construction, have now risen to key positions within the LIGO hierarchy. The LIGO community, once numbering in the dozens, has grown to include about a thousand collaborators located around the globe. Hundreds more are involved with gravity-wave detectors either constructed or planned in other countries, including Germany, Italy, Japan, and India. I wish I could have met them all but was grateful that many did open their doors to me as I swiftly worked on this update. For giving me their time, advice, and counsel this second go-round, I again thank GariLynn Billingsley, Mark Coles, Karsten Danzmann, Adalberto Giazotto, Peter Saulson, David Shoemaker, R. Tucker Stebbins, Kip Thorne, Rainer Weiss, and Michael Zucker. Also invaluable were the talks I had or e-mails I shared with Rana Adhikari, Stuart Anderson, Joseph Betzwieser, Kipp Cannon, Ryan de Rosa, Marco Drago, Anamaria Effler, Matthew Evans, Peter Fritschel, Valery Frolov, Joseph Giaime, Gabriela González, Shivaraj Kandhasamy, Joey Shapiro Key, Shane Larson, Nergis Mavalvala, Duncan Meacher, Cody Messick, William Parker, Deirdre Shoemaker, Amber Stuver, Keith Thorne (LIGO's *other* K. Thorne), Gary Traylor, and Virginia Trimble. And William Katzman at the Livingston Observatory helped me immensely as I made my second visit to the Louisiana detector.

I wish to give special thanks to the editors of *LIGO Magazine,* who provided a splendid narrative timeline—from discovery to announcement—on the first detection of a gravitational wave. It provided a ready-made path as I began my own explorations of those historic months. Several comments by LIGO collaborators within this book come from the magazine's issue on the detection with their permission.

Joseph Henry Press, a National Academies Press imprint that was my original publisher, is now (sadly) shut down, but both my NAP editor, Stephen Mautner, and my literary agent, Will Lippincott, lent very helpful hands in getting my rights transferred to Yale University Press for this updated edition. There I had the pleasure of working once again with both editor Joseph Calamia, whose physics expertise and sharp editorial eye added much to this book, and Laura Jones Dooley, who only enhances (never diminishes) a text with her superb copyediting.

My beloved Duffy is no longer with me, but there was a new loyal friend who remained by my side as I typed away in my home office. His name is Hubble—like Duffy, a bearded collie who is both a champ and a scamp. Last, my husband, Steve Lowe, deserves myriad hugs for his patience as I delayed many a dinner saying, "Just one more paragraph before I close for the day."

ACKNOWLEDGMENTS TO THE 2000 EDITION

I WAS FORMALLY INTRODUCED TO THE SCIENCE of gravitational waves in the 1980s. I was on assignment for *Science 85*, a magazine (now regrettably gone) that thrived on covering the cutting edge of science. Stanford University in California then housed the most advanced instrument yet constructed in the quest to detect gravitational waves—a hulking, supercooled bar of metal that resided in a cavernous room on the campus. This five-ton aluminum bar was the principal subject of my article on the continuing search for the elusive ripples in space-time. The best such a detector could hope to see, though, was a supernova explosion in our galaxy, which occurs only a couple of times each century.

However, while on the West Coast, almost as an afterthought, I decided to drop by Caltech to learn about another detection scheme that promised to see more sources in the long run but was described as being in its "infancy" compared to the bars: laser interferometry. Caltech housed a prototype on campus and during my visit I was shown a crude drawing of the full-blown observatory they hoped to construct one day in collaboration with MIT. The picture displayed two giant tubes that sprawled for miles over an imaginary desert plain. At the time, with federal budget cuts so prominent, I doubted I would ever see such an instrument built in my lifetime.

To my surprise and delight I was wrong. I use the word delight for good reason. At the time I began to write on astronomy and astrophysics, the various means of observing the universe's electromagnetic radiations were fairly well established. Astronomers had gotten a good start at examining the cosmos over the entire electromagnetic spectrum, from radio waves to visible light to gamma rays. I figured that the era had passed when a science writer could chronicle the advent of an entirely new astronomy, where the heavens were a blank slate ready to be filled in. But gravitational-wave astronomy, I came to realize, now offered me that opportunity.

For showing me the way, I must initially thank Peter Saulson, whom I first met in 1988 as I was writing a follow-up story on the field's progress. Over the intervening years, during which Peter moved from MIT to Syracuse University, he kept me abreast of the advancing technology and planted the notion to expand my magazine coverage into a book. His encouragement initiated the project; his sage advice and guidance followed me through to its completion. He and his wife Sarah have become cherished friends in the process.

When not involved with either interviewing or traveling—from Pasadena to Pisa—I could usually be found at the Science Library of the Massachusetts Institute of Technology by the banks of the Charles River. I would like to thank the many archivists and librarians there, who so patiently answered my questions and pointed me in the right direction toward the stacks.

By the end of my research, I had interviewed more than fifty scientists and engineers, either in person, over the phone, or via e-mail communications. My appreciation is extended to those who kindly agreed to review selected sections of my book, catching my errors and providing additional insights. They are John Armstrong, Peter Bender, GariLynn Billingsley, Philip Chapman, Karsten Danzmann, Sam Finn, Ron Dreyer, William Folkner, Robert Forward, Adalberto Giazotto, William Hamilton, Russell Hulse, Richard Isaacson, Albert Lazzarini, Fred Raab, Roland Schilling, Irwin Shapiro, Joseph Taylor, Kip Thorne, Tony Tyson, Stan Whitcomb, Clifford Will, and Michael Zucker. (Any errors that do remain are entirely of my own doing.) I am particularly grateful to Rainer Weiss, who in the midst of deadline pressures during LIGO's commissioning always welcomed me and found the time to answer my questions. Robert Naeye, a favorite editor of mine, also provided a keen editorial eye.

Thanks must also be given to those physicists and researchers who provided an essential historic perspective. They include John Wheeler, who took me on a delightful walk around Princeton University, where I got to see many of Einstein's old haunts, and Joseph Weber, who was a gracious host during my visit to the University of Maryland. For informative tours of the various detector sites, I thank Jennifer Logan at the Caltech prototype, Carlo Bradaschia at Virgo, Cecil Franklin at the Louisiana LIGO, and Otto Matherny at the Hanford LIGO. At Caltech, Donna Tomlinson, administrator extraordinaire, offered invaluable assistance in arranging my visit to the LIGO headquarters. I am equally grateful to the investigators who provided either key resource material or insightful discussions of their work. They are Barry Barish, Biplab Bhawal, Rolf Bork, Jordan Camp, Mark Coles, Douglas Cook, Robert Eisenstein, Jay Heefner, Jim Hough, Vicky Kalogera, Ken Libbrecht, Phil Lindquist, Walid Majid, Dale Ouimette, E. Sterl Phinney, Janeen Hazel Romie, R. John Sandeman, Gary Sanders, David Shoemaker, R. Tucker Stebbins, Serap Tilav, Wim van Amersfoort, Robbie Vogt, and Hiro Yamamoto.

Throughout the course of this project, friends and colleagues kept my spirits high by staying in touch, even as I would disappear for weeks on end. Their interest in my progress served as an incentive to keep moving forward. For this I thank Elizabeth and Goetz Eaton, Elizabeth Maggio and Ike Ghozeil, Tara and Paul McCabe, Suzanne Szescila and Jed Roberts, Fred Weber and Smita Srinivas, Linda and Steve Wohler, L. Cole Smith, Ellen Ruppel Shell, and Dale Worley.

Meanwhile, my goddaughter Skye McCole Bartusiak brought me smiles from afar with her varied entertainments. I am also grateful for the unwavering support of my mother, my brother, Chet, and his family, my sister, Jane Bailey, and her family, as well as Clifford, Eunice, and Bob Lowe, my husband's family. A special thank-you is extended to Duffy, my favorite neighborhood dog, who would energetically bark at my door to make sure I remembered it was important to get up from the computer, go outside, and walk around to smell the roses from time to time.

The reader would not be viewing this work at all, though, if it were not for the singular persistence of Stephen Mautner, executive editor of the Joseph Henry Press, whose powers of persuasion are formidable. He never faltered in his faith in this project and patiently waited nearly two years for me to sign on. (Thank you, Margaret Geller, for introducing us.) I have now discovered what a joy it is to work with a publishing house where effective communication of science is a priority and not just a sideline. Steve makes it all possible. Others at Joseph Henry that I must thank for their enthusiastic support behind the scenes are Barbara Kline Pope, Robin Pinnel, and Ann Merchant. Christine Hauser was an immense help in tracking down elusive photos, while copy editor Barbara Bodling skillfully honed my manuscript to a fine sheen.

Last, I must acknowledge my husband, Steve Lowe, whose love, encouragement, and support throughout these last two years allowed my ideas to reach fruition. More than that, I came to depend on his superb editorial judgment on matters scientific. Thank you, Steve, for being there.

EINSTEIN'S UNFINISHED SYMPHONY

Prelude

Ah, but a man's reach should exceed his grasp,
Or what's a heaven for?
              —Robert Browning, *Andrea del Sarto*, 1855

TO ARRIVE AT THE ASTRONOMICAL OBSERVATORY OF the twenty-first century, you must journey through America's Old South—Baton Rouge, Louisiana, to be precise. Coming off a plane, visitors are greeted by the pungent
smells wafting from a nearby petroleum refinery, a major state industry. As
you continue along Interstate 10, which partly follows the winding path of
the Mississippi River on its way to New Orleans and the Gulf of Mexico,
billboards hawk snuff, national catfish day, and Louisiana mud painting.
East of the state capital in plantation country along Interstate 12, the land
turns flat, the road arrow-straight like the unswerving grids on a sheet of
graph paper. The regularity is relieved only by sinuous chains of trees,
green rivers of foliage laden with Spanish moss, that occasionally meander
through the farmland.

Louisiana has some fourteen million acres of forests, so long-bed trucks
piled high with cut lumber are a familiar sight on its highways. Many of
these trucks originate about twenty-five miles east of Baton Rouge, in the
parish of Livingston, where a vast pine reserve is located. A few miles past
a deserted fireworks store, north along Highway 63, a modern sign announces the presence of LIGO (pronounced lie-go), which is operated by

the California Institute of Technology (Caltech) and the Massachusetts Institute of Technology (MIT). Those in the know realize that LIGO stands for Laser Interferometer Gravitational-Wave Observatory.

Nothing can be seen from the highway, though. You must first turn onto a country road and slowly drive past young-growth woodlands, some sections newly cut. Years ago longhorn cattle used to graze lazily along the roadway. After a mile's ride the observatory at last comes into view. It resembles a multistory warehouse, silver-gray in color, with blue and white trim. A prominent sign at the front gatehouse warns that a light will flash whenever there are "vibration-sensitive measurements in progress." Observatories are not a common sight in Louisiana, but establishing new technologies is a state tradition. In 1811 Edward Livingston, who would later become a US senator and the parish's namesake, helped Robert Fulton establish the first commercial steamboat operation on the Mississippi.

The vast room where LIGO's key instruments are mounted soars upward for more than thirty feet. The cross-shaped hall resembles an aircraft hangar or perhaps the transept and nave of a modern-day basilica. On two ends of the cross, at right angles to each other, are large, round ports. Attached to each of these two openings is a long tube that extends out into the countryside for two and a half miles (four kilometers). Each four feet wide, the two tubes resemble oil pipelines and are roomy enough for a crablike walk should the need arise. To accommodate the pipes, the pine forest has been leveled. A huge, raw L has been carved into the woody terrain to make room for these lengthy metal arms. One long swathe stretches to the southeast, the other to the southwest. Alongside each arm a roadway has been constructed about eight feet above the Louisiana floodplain. The dirt for the roadway was dug up on site, creating two water-filled canals, each running parallel to a tube. An alligator, fed doughnuts by the construction crew, even adopted one of the borrow pits as his home. The pipes cannot be seen directly. Concrete covers, six inches thick, protect them from the wind and rain, as well as any stray bullet that might pass by during hunting season. A hit could be devastating, for the pipes are as empty of air as is the vacuum of space.

"Here's our two-and-a-half-mile-long hole in the atmosphere," says Mark Coles, sweeping his hand outward with pride. A big and friendly

LIGO in Livingston, Louisiana. *Courtesy of Caltech/MIT/LIGO Laboratory.*

man, Coles, now a senior adviser at the National Science Foundation, moved from Caltech in 1998 to become the first director of the Livingston observatory. He took to the Cajun lifestyle with ease, even coming up with a bit of French for the observatory's slogan, which was once displayed on T-shirts sold in the reception area: *Laissez les bonnes ondes rouler* (Let the good waves roll!). The waves LIGO researchers seek are those of gravitational radiation—gravity waves in common parlance. (In physics, the term *gravity wave* has long been used officially to refer to another phenomenon—an atmospheric disturbance in which a wave originates from the relative buoyancy of gases of different density. For simplicity's sake, I use the term here as a synonym for *gravitational wave,* as researchers themselves often do when speaking casually. To quote Shakespeare, the term's succinctness comes "trippingly on the tongue . . . that may give it smoothness.")

Electromagnetic waves, be they visible light, radio, or infrared, are released by individual atoms and electrons and generally reveal a star or other celestial object's physical condition—how hot it is, how old it is, or what it is made of. Gravity waves convey different information. They tell us about the overall motions of massive celestial objects. They are literally quakes in space-time that emanate from the most violent events the universe has to offer—a once-blazing sun burning out and going supernova, the dizzying spins of neutron stars, the cagey dance of two black holes pirouetting around each other, approaching closer and closer until they merge. Gravity waves tell us how large amounts of matter move, twirl, and collide throughout the universe. Eventually, this new method of examining the cosmos may even record the remnant rumble of the first nanosecond of creation, what remains of the awesome space-time jolt of the Big Bang itself. At the dedication of the Livingston observatory in 1999, Rita Colwell, then director of the National Science Foundation, noted that they were "breaking a bottle of champagne over the figurative bow of a modern-day galleon—a gravity-wave observatory that may ultimately take us farther back in time than we've ever been." So compelling is this quest that scientists have pushed the envelope of technology (many times over) to detect these subtle tremors.

Inside LIGO's main hall the ambiance is almost reverential, akin to the response one might feel inside a darkened telescope dome, the mirror aimed for the star-dappled heavens. But this astronomical venture is vastly different. Within this domain there are no literal windows to spy on the universe. A gravitational-wave observatory is securely planted on cosmos firma. It is glued to space-time, awaiting its faint rumbles, vibrations first predicted by Albert Einstein more than a hundred years ago.

"The worship of Einstein, it's the only reason we're here, if you want to know the truth," says Rainer Weiss of MIT. "There was this incredible genius in our midst, in our own lifetime. The average person knows that he was an important guy. If you go to Congress and tell them you're going to try to show that Heisenberg's uncertainty principle is not quite right, you run into blank stares. But if you say you're measuring something that's proving or disproving Einstein's theory, then all sorts of doors open. There's a mystique."

Einstein indeed stands like a giant amid the pantheon of scientific figures of the twentieth century. His ideas unleashed a revolution whose changes are still being felt into this new century. What he did was to take our commonplace notions of space and time and completely alter their definitions. The evolution of physics in many ways traces these changing conceptions of space and time. And each adjustment in our worldview—our cosmic frame of reference—brought with it an accompanying change in the physics of handling that new framework. It is the start of every physicist's musings: an attempt to track an object's motion in space and time. These intangibles are given names, such as "miles" and "seconds," and ever since Sir Isaac Newton allegedly analyzed that falling apple, scientists believed that they finally understood the true meanings of those descriptive words.

But they didn't. Einstein stepped in to shatter that confidence. With his general theory of relativity, he showed us that space and time are not separate entities but rather eternally linked, producing the force we know as gravity in its interaction with matter. That is why Einstein initiated a revolution. He taught us that space and time are not mere definitions useful for measurements. Instead, these two entities are fused to form a specific object, known as space-time, whose geometric shape is determined by the matter around it. According to general relativity, massive bodies, such as stars, dimple the space-time around them (much the way a bowling ball placed on a trampoline would create a depression). Planets and comets are then attracted to the star simply because they are following the curved space-time highway carved out by the star.

To the practiced eye, Einstein's equations stand as the quintessence of mathematical beauty. When it was introduced in 1915, general relativity was hailed as a momentous conceptual achievement. But for a long time the theory had little practical importance. Although the scientific community embraced general relativity—and recognized Einstein for his genius—the evidence for it was largely aesthetic, because the tests possible in the early twentieth century were few: a tiny shift observed in the planet Mercury's orbit and evidence of starlight bending around the Sun due to the indentation of space around the orb's enormous mass. There was a perfectly good reason that experimental tests lagged behind theorists' contemplation of relativity's effects. In either describing the motion of a ball falling to the ground or sending a spacecraft to the Moon, Newton's laws of gravity are still adequate. General relativity is far more subtle, best observed when gravitational fields attain monstrous strengths, making its effects more prominent. The nature in which we reside, which flows so nicely according to Newton's laws, is actually a special case, where energies and velocities are relatively low. It is when energies get higher that nature begins to act as if it's a character in *Alice's Adventures in Wonderland*—the universe begins to look "curiouser and curiouser." It's a counterintuitive world, where lengths can contract, time can speed up or slow down, and matter can disappear in a wink down a space-time well.

But in Einstein's day the universe was deemed far tamer than that. After the first few tests were completed—elevating Einstein to the pinnacle of celebrity—general relativity became largely a theoretical curiosity, admired by all but exiled into the hinterland of physics. "General relativists," says Clifford Will, a relativist himself, "had the reputation of residing in intellectual ivory towers, confining themselves to abstruse calculations of formidable complexity."

But like some theoretical Rip Van Winkle, general relativity gradually revived after decades of neglect, especially as astronomers came to discover a host of intriguing celestial objects, such as pulsars, quasars, and black holes, that could be understood only through the physics of general relativity. Neutron stars, gravitational lenses, inflationary universes—all must be studied with Einstein's vision in mind. At the same time, advancing technology offered physicists new and ingenious means to test relativity's quirky and subtle effects with unprecedented accuracy—and not just in the laboratory. Using planetary probes, radio telescopes, and spaceborne clocks as their tools, investigators have checked Einstein's hypotheses with uncanny degrees of precision. The entire solar system now serves as general relativity's laboratory. In the words of general relativity experts Charles Misner, Kip Thorne, and John Wheeler, from their book *Gravitation,* a veritable bible for workers in the field, "General relativity is no longer a theorist's Paradise and an experimentalist's Hell."

For more than a century the theory has held up to every experimental test, which is not only a triumph for Einstein—a celebration of his accomplishment—but also a further example of nature's ability to inspire awe and fascination in the way its rules follow such a precise mathematical blueprint. Furthermore, there are practical reasons why the theory is no longer relegated to the ivory tower. General relativity now has a real impact on our everyday lives. To operate properly the satellites of the Global Positioning System (GPS)—used regularly by hikers, mariners, commuters, and soldiers to keep track of their locations—requires corrections of Einsteinian precision, revisions unaccounted for by Newton's cruder take on gravity. And astronomers who electronically link radio telescopes on several continents, creating a telescope as big as the world for their observations, also demand such accuracy.

And yet the story of general relativity remained incomplete for quite some time. An unsettled question resided within general relativity, a major prediction that awaited many decades for direct confirmation: gravitational waves. To understand this phenomenon, imagine one of the most powerful events the universe has to offer—a supernova, the spectacular explosion of a star. More than 160,000 years ago, at a time when woolly mammoths were walking the Asian plains, a brilliant blue-white star now known as Sanduleak −69° 202 exploded in the Large Magellanic Cloud, a prominent celestial landmark in the southern sky. Not until the winter of 1987, though, did a wave of neutrino particles and electromagnetic radiation shot from that dying star reach Earth. And when that blitz arrived, an arsenal of observatories around the world focused their instruments on the flickers of light and energy that represented the star's long-ago death throes. It was the first time that astronomers were able to observe a supernova in our local galactic neighborhood since the invention of the telescope.

But Einstein's theory suggests that, when Sanduleak −69° 202 blew, it also sent out waves of gravitational energy, a spacequake surging through the cosmos at the speed of light. A fraction of a second before the detonation, the core of the star had been suddenly compressed into a compact ball some ten miles wide, an incredibly dense mass in which a thimbleful of matter weighs up to five hundred million tons (roughly the combined weight of all of humanity). This is the birth of what astronomers call a neutron star. Jolted by such a colossal stellar collapse, space itself was likely shaken—and shaken hard. The resulting ripples would have rushed from the dying star as if a giant cosmic pebble had been dropped into a space-time pond. These gravitational waves, though growing ever weaker as they spread from the stellar explosion, would have squeezed and stretched the very fabric of space-time itself. Upon reaching Earth they would have passed right through, compressing and expanding, ever so minutely, the planet and all the mountains, buildings, and people in their wake.

As shown by the example of the Sanduleak star, gravity waves are generated whenever space is fiercely disturbed. These waves are not really traveling through space, say in the manner a light wave propagates; rather, they are an agitation of space itself, an effect that can serve as a powerful probe. Light beams are continually absorbed by cosmic debris—stars, gas clouds, and microscopic dust particles—as they roam the universe. Gravity waves, by contrast, travel right through such obstacles freely, since they interact with matter so weakly. Thus, the gravity-wave sky is vastly different than the one viewed by astronomers with telescopes. More than offering an additional window on space, gravity waves provide a radically new perception. In addition, they at last offered proof of Einstein's momentous mental achievement—their very existence demonstrates, in a firm and vivid way, that space-time is indeed a physical entity in its own right.

A few pioneering investigators in both the United States and Europe claimed to have detected the Sanduleak star's faint space-time rumble, but those claims were erroneous. The missed opportunity, however, made the early developers of gravitational-wave astronomy more determined than ever not to be caught off guard when another wave rolled by.

LIGO itself consists of two observatories, one in Louisiana and a twin in Washington State, that operate in concert with each other. But they are not alone. Similar instruments of varying size have been built in Italy (the Virgo detector) and Germany (GEO600), and other detectors are being commissioned in Japan and planned for India.

No one doubted that gravitational waves existed, for there was already powerful evidence that such waves were real. Two tiny neutron stars—supernova remnants—in our galaxy have been observed rapidly orbiting each other. They are drawing closer and closer together. The rate of their orbital decay, about a few yards per year, is just the change expected if this binary pair is losing energy in the form of gravitational waves. Though the proof was indirect (and the waves themselves were too weak to measure), it encouraged the gravity-wave community that sources were available.

In the early 1600s, Galileo Galilei, then a savvy professor of mathematics at the University of Padua in Italy, pointed a newfangled instrument called a telescope at the nighttime sky and revealed a universe with more richness and complexity than previous observers ever dared to contemplate. The ancients had said that the heavens were perfect and unvarying, but Galileo discovered spots on the Sun and jagged mountains and craters on the Moon. With bigger telescopes came bigger revelations. We came to see that the Milky Way was not alone but one of many other "island universes" inhabiting space. More than that, all of these galaxies were rushing outward, caught up in the expansion of space-time. And as astronomers were able to extend their eyesight beyond the visible light spectrum and detect additional electromagnetic "colors," such as radio waves, infrared waves, and X-rays, the heavens underwent a complete renovation. Long pictured as a rather tranquil abode, filled with well-behaved stars and elegant spiraling galaxies, the cosmos was transformed into a realm of extraordinary vigor and violence. Arrays of radio telescopes, aimed toward distant realms of the visible universe, have observed young, luminous galaxies called quasars spewing the energy of a trillion suns out of a space no larger than our solar system. Focusing closer in, within our own stellar neighborhood, these same radio telescopes watched how neutron stars—city-sized balls of pure nuclear matter, the collapsed remnants of massive stars—spin dozens of times each second. Meanwhile, X-ray telescopes discovered huge amounts of X-ray-emitting gas, unobservable with optical telescopes, hovering around large clusters of galaxies. The invisible became visible.

The twenty-first century is seeing the sky remade once again. It is happening now that astronomers at last detected gravity waves in 2015. These ripples in space-time are detected neither with the eye nor as the image of a star or galaxy on an electronic display, not in the same way that visible light waves, radio waves, or X-rays are distinguished. Each gravity wave that passes by Earth is, in a way, felt—perceived variously as a delicate vibration, a vibrant boom, or even a low-key cosmic rumble. As a result, astronomy will never be the same again. It's as if, in studying the sky through the centuries, we've been watching a silent movie—pictures only. But gravity waves are now adding sound to our cosmic senses, since their frequency can fall into the audio range. They are *not* sound waves; rather, much like the ridged grooves in a vinyl record, the signal recorded by a gravity wave detector can be played back as a sound. In this way the silent universe has been turned into a talkie, a motion picture with a soundtrack, one in which we hear the thunder of colliding black holes or the whoosh of collapsing stars. When the LIGO observatories detected a gravity wave on September 14, 2015—a historic occasion for general relativity—we heard the first notes of Einstein's symphony, a hyperfast chirp.

A gravitational-wave telescope essentially acts like a geological seismometer, but a seismometer that is placed on the fabric of space-time to register its temblors. The oldest detectors were designed as car-sized, cylindrical metal bars, capable of ringing like bells if a sizable gravity wave had passed through them. The newest instruments, such as LIGO, involve a set of suspended weights that sway as the peaks and troughs of the traversing gravitational wave alternately squeeze and stretch the space between the masses (although these movements are exquisitely tiny, the displacement being a thousand or more times smaller than the width of an atomic nucleus). Once various observatories around the globe are able to work together consistently, each will act as a surveyor's stake, enabling gravity wave astronomers to triangulate the source on the sky and better determine where the wave originated. A gravity-wave signal can be regular or erratic, unceasing or sporadic—in essence, a cosmic symphony of beats. And gravitational-wave astronomers will be translating these syncopated rhythms—the whines, the bursts, the random roars—into a new map of the heavens, a clandestine cosmos once impossible to see.

This entire endeavor began very modestly in the 1960s as one man's quixotic quest. At the University of Maryland, physicist Joseph Weber cleverly devised the first scheme to trap a gravitational wave and reported a detection in 1969. Inspired by Weber's insight, others quickly joined the campaign. Gravitational-wave detectors were erected around the world. In the end, Weber's sighting was never confirmed, but that didn't deter the newcomers to gravitational-wave physics from continuing the search. They were energized by the technological challenges of the problem. Weber triggered a movement whose momentum has never diminished to this day. A host of specialists—in optics, lasers, materials science, general relativity, and vacuum technology—came together to produce the most complex instruments ever devised for an astronomical investigation. Many in the astronomy and physics communities waged a strong campaign against the endeavor at first, declaring that the money would have been better spent on

surer scientific pursuits. LIGO's technological reach, they loudly proclaimed, was exceeding its grasp. But the potential of the science—not to
mention strong politicking—finally overrode those concerns. As a result,
gravitational-wave researchers are not just carrying out an experiment, they
are founding a field. They are now operating the most innovative tool with
which to explore the heavens. The questions they are asking stretch back to
Aristotle, and answers are now arriving. These efforts were the culmina-
tion of a centuries-long pursuit—nothing less than the ambition to unravel
the enigma of space and time.

Space in G Flat

WE TALK ABOUT SPACE SO READILY AND easily: "There's no space for an office in this apartment" or "Give me some space, man." The concept of space seems self-apparent to the casual observer. But on deeper reflection its true nature remains elusive. "As a rule, people differentiate between matter, space, and time. Matter is what exists in space and endures through time. But this does not tell us what space is. . . . It is matter which we see, touch, and hear, which causes sensations to arise within us," the British philosopher Ian Hinckfuss once noted. Space is perceived yet not felt, observed, or heard. So what, after all, is space?

Perceiving the bounds of space was probably one of *Homo sapiens*'s earliest accomplishments. Space was first and foremost an orientation among familiar objects, determining the effort needed to reach a nearby river, rock, or tree. A sense of space may even have preceded a perception of time, because we describe moments of time as "short" or "long," words that usually describe spatial categories. And with the emergence of agriculture came the need to measure space exactly for practical considerations, such as planting a field or digging an irrigation ditch.

From these humble beginnings arose a more esoteric contemplation of space. In ancient Greece philosophers developed the concept of the void—the *pneuma apeiron*—a vacancy that allowed for a separation of things. Democritus, father of the atomic theory, required this void—a nothingness bereft of matter. In order for his atomic theory to work, space had to be an empty extension, which allowed his bits of matter—his atoms—to move about.

Such discussions soon extended to thinking about space in the abstract. A Greek philosopher named Archytas asked what would happen if you journeyed to the end of the world and stretched out your hand. Would your hand be stopped by the boundaries of space? Lucretius, Democritus's pupil, answered no, and he had an interesting proof for space being unbounded. He asked: Suppose a man runs forward to the very edge of the world's borders and throws a winged javelin. Unable to conceive that anything could get in the javelin's way, Lucretius concluded that the universe must stretch on and on without end. Aristotle, on the other hand, took the opposite stand. He stated that it was "clear that there is neither place nor void nor time beyond the heaven." For him the universe was finite. If a stone falls to the Earth to find its natural place at the center of the universe, he argued, then fire moving upward in the opposite direction must also face a limit. In Aristotle's physics, upward and downward motions had to be balanced. Moreover, any objects in the farthest reaches of an infinite universe, forced to rotate about a motionless Earth, would end up traveling at infinite velocities, a situation that Aristotle considered patently absurd.

These fierce intellectual debates concerning space continued for centuries, and by medieval times theological concerns often prejudiced the arguments. To think of an immovable void, as outlined by the Greek atomists, was to imagine that God created something He could not budge, which challenged His omnipotence. Consequently, any thought of a stationary emptiness was deemed heretical and therefore shunned. But as natural philosophers, starting as early as the fourteenth century, began to consider kinematics, the motion of objects, it became necessary to contemplate some kind of fixed space. They needed to imagine a special, motionless container in order to understand such physical concepts as velocity and acceleration. One man in particular would change the landscape of science in making this assumption as he searched for the mathematical rules by which motion could be predicted. That man was Sir Isaac Newton.

The dreaded Black Death appeared in London in 1665 during the reign of Charles II. With the plague spreading northward to the university town of Cambridge, Newton fled that summer to his childhood manor home, Woolsthorpe, in Lincolnshire. In that rural setting the young scholar continued to work intensely on his studies. Still in his early twenties, he was laying down the mathematical and physical foundations of his most

A young Isaac Newton in 1677.
Wikimedia Commons.

important ideas, which had been germinating throughout his college years: the theory of color, the construction of the calculus, and, most important, the laws of gravitation. Possibly inspired by that fabled apple falling in his country garden, he began to reflect on the tendency of bodies to fall toward the Earth at a set acceleration. He wondered if the same force acting on an apple extended to the Moon. A brilliant mathematician, largely self-taught, he computed that the Moon did appear to be falling toward the Earth—its path becoming curved—by a pull that diminished outward by about the square of the distance. But these early computations were not as yet refined, so Newton put the problem aside.

He returned to Cambridge in 1667, at the age of twenty-four, and within two years became Lucasian Professor of Mathematics, a high honor at the university. Secretive, obsessive, and fearful of exposing his work to criticism—a man chockful of neuroses—Newton let many of his early revolutionary thoughts go unpublished. It was not until 1684, sparked by the

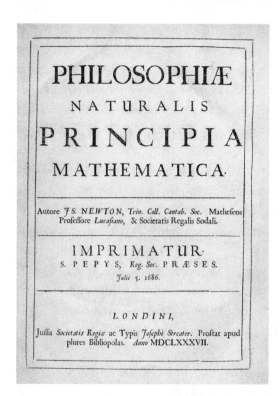

PHILOSOPHIÆ

NATURALIS

PRINCIPIA

MATHEMATICA·

Autore *JS. NEWTON,* Trin. Coll. Cantab. Soc. Mathefeos
Profeffore *Lucafiano,* & Societatis Regalis Sodali.

IMPRIMATUR·
S. PEPYS, Reg. Soc. PRÆSES.
Julii 5. *1686.*

LONDINI,

Juffu *Societatis Regiæ* ac Typis *Jofephi Streater.* Proftat apud
plures Bibliopolas. *Anno* MDCLXXXVII.

The title page of Newton's
monumental work on
gravity. *Wikimedia Commons.*

questions and persistent prodding of Edmond Halley (later to gain comet
fame), that Newton was at last convinced to write his masterpiece on gravitation, *Philosophiae naturalis principia mathematica* (Mathematical principles of
natural philosophy). He abandoned his work on alchemy, his most recent
fascination, and applied his legendary power of concentration completely on
the *Principia* (as it is most familiarly known) for nearly two years.

The *Principia*, published in 1687, deals with both gravity and the mechanics of motion. Forces in nature, Newton declared, are not needed to keep
things moving (as Aristotle argued); rather, forces change motion and in pre-
dictable ways. Newton clarified what Galileo had begun to infer from experimental tests: an object in motion does not naturally come to a stop; instead, it
will remain in motion unless altered by an outside force, such as friction. The
effect of a force is to get an object moving, to stop it, or to change its direction.

And when it comes to gravity, Newton revealed that the strength of
the gravitational attraction between two objects depends on two things: the
total amount of matter in each object and the distance between them. The

greater the mass of each object, the stronger the pull; conversely, the larger the separation between the two masses, the weaker the attraction. Or as Newton put it, two objects exert a gravitational force on each other that is in direct proportion to their masses and in inverse proportion to the square of their distance. That is, double the distance between two objects and the force between them is reduced to one-fourth of its original strength. Triple the distance and the force diminishes to one-ninth. He deduced the exact equations for determining those motions and, in doing so, discovered that nature uses a mathematical treatise as its playbook. And this was true whether the object was an apple or a celestial moon.

As if with one monumental stroke of the pen, Newton thus established that the same physical laws describe motions *everywhere*—both in the heavens and on Earth. Before this insight, philosophers generally believed that the heavens were distinctly different from the domain of humanity. Earthly things were mortal—subject to change and transition—while the stars and planets were eternal and incorruptible. But with Newton's new laws the cosmos and terra firma were blissfully wedded. An all-encompassing set of mathematical rules could now explain events in both domains: the ocean tides, the motion of comets and planets, as well as the projectile paths of cannon balls. All these phenomena could be tracked with the same clock-like precision. So great was this achievement that Newton was the first person in England to be knighted for his scientific work.

Skydivers and bungee jumpers, plummeting toward the ground, have great respect for the force of gravity pulling them downward. Gravity is also the ruling force in determining the universe's evolution and its grand structure. Yet gravity is the weakest force in the cosmos. It seems paradoxical. A toy magnet can easily pick up a paper clip against the gravitational pull of the entire Earth. Or take two protons sitting next to each other. The gravitational force between them is a trillion, trillion, trillion times weaker than the electrostatic force pushing on them. Gravity gains collective strength only when masses accumulate and exert their effect over larger and larger distances. In this way gravity comes to control the motions of planets, stars, and galaxies.

In order to devise his new rules on motion and gravity, Newton realized that he first needed to establish a framework. Take Newton's first law of motion. A body either remains at rest or in continuous uniform motion (traveling in a straight line at a constant speed), unless an outside force causes that state to change. But at rest in relation to what? Or in motion to or away from what location? As soon as one talks of "motion," one must establish a home base. Think of a child reading a book in a moving car. To someone on the curb watching the car whiz by, the book is moving fast. To the child inside, it is perfectly still. Newton's critical choice was to establish a single reference frame—the universe at large. Space itself became his motionless laboratory: flat, penetrable, yet forever the same.

He was not the first to think this way—Galileo, for one, posited a continuous three-dimensional void—but Newton made it an integral component of the Western canon. "Absolute space in its own nature, without relation to anything external, remains always similar and immovable," he authoritatively stated. Space was at rest, and everything else in the universe moved with respect to that. To Newton, space was an empty vessel. You were either at rest or in motion with regard to this container. Positions, distances, and velocities were all measured in regard to this fixed space. Only by establishing this framework—this unchanging cosmic landscape—could his equations work.

Measuring motions in this absolute space also required a universal clock, which ticked off identical seconds for all the inhabitants of the cosmos. Events everywhere, from one end of the universe to the other, were in step with the ticks of this grand cosmic timepiece, no matter what their speed or position. A clock sitting at the edge of the universe or zipping about the cosmos at high speed would register the same passage of time, identical minutes and identical seconds, as an earthbound clock. This meant that two cosmic observers, perched on opposite sides of the universe, could synchronize their watches instantaneously. Moreover, "the flowing of absolute time is not liable to any chance," said Newton in the *Principia*. His clock was never affected by the events going on around it. Like some cosmic Big Ben, time stood aloof, as solar systems formed and moons orbited planets in this vast universe of ours.

Newton's law of gravity, brilliant in its ability to predict the future paths of planets, did have an Achilles' heel. It provided no explanation for the mechanism underlying gravity. There was no medium or physical means that was pushing or pulling the planets and other objects around. Newton's tendrils of gravitational force simply appeared to act instantaneously over vast distances, as if by magic. This feat appeared more resonant with the occult. As one wry critic of the time noted, "Newton calculated everything and explained nothing." For some, his inability to establish the source of gravity's power was tantamount to bad science. Newton was aware of these difficulties and lamented that one body acting on another through a vacuum, "without the mediation of anything else, by and through which their action and force may be conveyed from one to another, is to me so great an absurdity, that I believe no man who has in philosophical matters a competent faculty of thinking, can ever fall into it." But his decision to stand by his laws was a practical one—they worked! His equations allowed successful predictions to be made when dealing with motion or gravity. As Einstein said in an imaginary talk with Newton, "You found the only way which, in your age, was just about possible for a man of highest thought and creative power." Newton's laws became ingrained into the very fabric of physics simply because they came up with the right answers.

Newton's conception of an absolute space and absolute time influenced the course of physics for some two hundred years, but it was not universally accepted. There were critics who raised their voices loudly. The most notable were the British philosopher George Berkeley and the German diplomat and mathematician Gottfried Wilhelm Leibniz, who was Newton's archrival for his claim to have invented the calculus first. To these two men, space and time were not fixed entities at all. Leibniz declared that "space and time are orders of things, and not things." Space and time could only be defined with regard to their relation with objects of matter. For Muslim philosophers, who had developed similar theories, this avoided the question, "Where was God before the creation?" The answer was simply that there was no place. Space did not exist until the creation of matter.

Toward the end of his life, Newton found solace for his worries about absolute space and time in a religious explanation: "[God] endures forever, and is everywhere present; and by existing always and everywhere, he constitutes duration and space." For the Greek philosophers, mathematics had largely been an aesthetic experience. Newton changed that attitude by demonstrating, with his powerful law of gravitation, that mathematics offered a road to discovery. He transformed mathematical law into physical law, rules by which the operations of nature—planetary motions, the propagation of light, mechanics—could be calculated. Eventually belief in the infallibility of these laws grew so strong that when Newton's equations appeared to fail (as in the case of explaining the orbital motions of the planet Uranus), it was immediately assumed that the equations were still valid and that an unseen planet was lurking beyond Uranus to account for the discrepancy. In that particular case, such resolute faith paid off handsomely. Adherence to Newton led to the discovery of the planet Neptune in 1846. It was difficult for Newton's critics to fight such success.

In his mathematical choices Newton assumed that space was "Euclidean"— that is, holding all the properties defined by the famous Greek geometer Euclid in the third century B.C.E. Although the rudiments of geometry had been fashioned in Egypt along the banks of the Nile River—knowledge gained by the pharaoh's surveyors, the *harpedonaptai* or rope-stretchers—those rules did not turn into a rigorous discipline until they traveled to Greece. The Greek philosophers saw in geometry a vibrant set of truths that could be arrived at through logic alone. Geometry was proof that knowledge of the physical world could be gleaned through pure reason. So revered was geometry that when Plato established his Academy it was said that the sign over the door announced, "Let no one enter here unless he knows geometry." And it was Euclid who stood at the zenith of this movement. Around 300 B.C.E. he wrote *The Elements,* which summarized all the geometrical knowledge then known into a concise set of axioms and postulates. This historic work served as the basis for mathematical thought for the next two thousand years.

In this masterpiece Euclid defined space as flat, which is exactly the world we perceive and measure around us when confined to the ground. He also made a list of all the geometric ideas we take for granted—for example, that a straight line can be drawn between any two points or that all right angles are equal to one another. These were self-evident truths. His fifth postulate considers a line and any point not on that line. According to the ancient Greek geometer, there is one—*and only one*—line that can be drawn through that point that is parallel to the original line. The two lines, like two parallel railroad tracks, will never meet. To our senses there appears to be no other possible configuration.

While it seems self-apparent that parallel lines will never intersect, later mathematicians began to examine this particular axiom in more depth. Rather than assuming it was a given, they wondered whether the rule could be derived directly from Euclid's other axioms. They wanted to prove it explicitly, rather than just state it as true.

In order to test a postulate, one tried-and-true mathematical trick is to assume that it is false, then see what happens. That's precisely what a Jesuit priest named Girolamo Saccheri did in 1733. He assumed that the parallel axiom was false. And in doing so, he demonstrated that it would lead only to absurdities—hence, the name of this technique, *reductio ad absurdum*. Saccheri discovered that he could get *more* than one line through a point to be parallel to a given line. Figuring that this was clearly ridiculous, he concluded that he had proven what he had set out to do—show that the axiom was clearly true, as stated so elegantly by Euclid. What Saccheri failed to see is that he had accidentally stumbled onto a whole new geometry. It took nearly a century before this advance was fully recognized.

Not until 1816, after years of thinking about the problem, did another mathematician arrive at the same insight. And he too backed off, but this time out of fear of ridicule. The great German polymath Carl Friedrich Gauss uncovered the same absurdities that Saccheri arrived at but didn't reject them outright; he simply knew that such challenges to the great Euclid would be considered heresy. Gauss was also a perfectionist who kept much of his work to himself. He was terribly reluctant to publish any idea until he had polished its proof to a fine sheen. It's not surprising that his personal seal, a tree with sparse fruit, bore the motto *Pauca sed matura* (Few but ripe). As a result, Gauss never officially reported his findings during his lifetime (although he did discuss his new geometry in private correspondence with colleagues). A reclusive man unwilling to start a public dispute that would disrupt his peace of mind, Gauss carefully guarded his secret, fearing as he put it, "the clamor and cry of the blockheads" over questioning mathematics' sacred gospel. Euclid's framework, sturdy for centuries, served as the very foundation of mathematics.

Though Gauss didn't publish his results, he did think hard about them. Realizing that non-Euclidean geometries were a possibility (at least on paper), he gradually began to wonder whether a non-Euclidean geometry might describe true physical space. Perhaps space wasn't flat, as Newton assumed, but rather curved. His musings were amplified by a practical concern. The government commissioned him in the 1820s to conduct a geodetic survey of the region around the city of Göttingen in Hannover. This endeavor heightened the thoughts he was already having about curved space. Spatial curvature, he realized, need not be restricted to two dimensions, as Euclid and Newton had us believe. In an 1824 letter to Ferdinand Karl Schweikart, a professor of law and a geometer as well, Gauss bravely mentioned that space itself, in all its three dimensions, might be curved or "anti-Euclidean," as he called it. "Indeed," he wrote, "I have . . . from time to time in jest expressed the desire that Euclidean geometry would not be correct." He may even have tested this hypothesis during his geodetic work. Using light rays shining from peak to peak in the Harz Mountains, Gauss surveyed a triangle of pure space formed by three mountains, the Hohenhagen, the Brocken, and the Inselberg. By his metric figuring, the sides of the triangle measured 69, 85, and 107 kilometers (43, 53, and 66 miles, respectively). No deviation from flatness, however, was detected.

While Gauss kept mum from his faculty post at the University of Göttingen, others were more open as they explored this new geometric terrain. Between 1829 and 1832, two mathematicians independently published papers stating that it was indeed possible to have geometries that disobeyed Euclid. One proof was done by the Russian mathematician Nikolai Lobachevsky, the other by the Austro-Hungarian János Bolyai, allegedly the best swordsman and dancer in the Austrian imperial army in his day. Like Saccheri a century earlier, Lobachevsky and Bolyai had asked what would happen if Euclid's fifth postulate were wrong. If it were in error, what type of mathematics arises? What if it is assumed that an infinite number of straight lines can be drawn through a point near a given line without any of the lines intersecting? In this way the two mathematicians came to describe a space of *negative* curvature.

"From nothing I have created another entirely new world," wrote Bolyai to his father, who had struggled with the fifth postulate himself when he was a student friend of Gauss. Bolyai's new world can be visualized by imagining a triangle drawn on the surface of a saddle. The triangle, with its sides curved inward, would appear a bit shrunken. So the sum of its angles is not 180 degrees, as authoritatively stated in high school textbooks as the standard Euclidean answer. Instead, it is less than that. Being concave, the saddle's surface also allows many lines, which never meet, to be drawn through a point near a given line. Lobachevsky called this new system his "imaginary geometry."

Like Gauss, Lobachevsky also thought of the possibility that three-dimensional space might be curved but figured that distances far longer than the spans between alpine mountains would be needed to test such a radical idea. He suggested conducting measurements using distant stars as surveyor's stakes to see whether a triangle extended into space added up to less than 180 degrees. When astronomers at last carried out such celestial surveys in the mid-nineteenth century, no change from flatness was discovered. Hence, it was assumed that Euclid's rules continued to reign supreme throughout the universe.

Meanwhile, Gauss's fascination with the new geometry was passed on to a brilliant student at the University of Göttingen, Bernhard Riemann, who developed another non-Euclidean geometry altogether. The timid twenty-seven-year-old revealed his new construct during a trial lecture he gave while seeking appointment as a *Privatdozent* (lecturer) at the university in 1854. In his talk, prepared in only seven weeks and later described as a high point in the history of mathematics, Riemann introduced a geometry in which *no* parallel lines can be drawn through a point near a given line. He was dealing with a space of *positive* curvature, best typified by the surface of a sphere. Here, the shortest distance between two points is not a straight line in the Euclidean sense; instead, the shortest path is an arc, a segment of a great circle that encompasses the entire sphere.

Consider two adjacent lines at Earth's equator aligned directly north to south. Locally, they seem as parallel as they can be. Yet extend these lines around the world and they eventually cross and meet at the poles. Consequently, there are no parallel lines in this special kind of geometry. A triangle on such a curved surface would look inflated; the sum of its angles would be more than 180 degrees. Like Gauss, Lobachevsky, and Bolyai before him, Riemann was discovering that a mathematician can imagine myriad geometric worlds. Euclid did not corner the market at all.

Infamous for his stern and critical demeanor, Gauss displayed a rare enthusiasm at the end of Riemann's presentation. He was perhaps the only

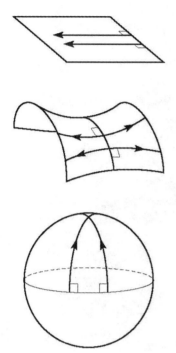

Three different geometries: flat space (*top*);
negatively curved space (*middle*); and positively
curved space (*bottom*).

one in the audience that day who recognized that Riemann had surpassed
his predecessors by extending non-Euclidean geometry much, much farther. During his lecture, Riemann proceeded to generalize the geometry of
curved spaces to higher dimensions—spaces involving four, five, and even
more dimensions. Although these manipulations may have appeared at the
time to be no more than a mathematical game, this work later proved invaluable when Einstein faced the awesome task of developing his general
theory of relativity. Riemann fashioned the tools that allowed Einstein to
envision a completely different view of space and time. Riemann served as
a vanguard for the Einsteinian revolution to come.

Riemann's prescience was demonstrated when he dared to suggest that
the true nature of space would be found not in ancient manuscripts from
Greece but rather from physical experience. He once imagined that the universe might close in on itself, forming a sort of four-dimensional ball. Such
a curvature would be noticed only over great distances, hence our common

Bernhard Riemann three years before his death in 1866. The
mathematics he developed later allowed Einstein to construct his
general theory of relativity. *Wikimedia Commons.*

experience of seeing our local universe as flat. Of more interest, Riemann
went on to consider whether the structure of space was somehow molded
by the presence of matter, creating what he called a *metrical* field, akin to the
electromagnetic field that James Clerk Maxwell postulated in the 1860s. It
was a prophetic vision, but Riemann spoke too soon. Physics was not yet
ready to give up on its pleasant Newtonian world, a world of absolute and
rigid space, unvarying and unchanging. The idea that space might display a
distinct geometry elicited fury in certain philosophers of that era. Space was
still considered an empty vessel devoid of physical properties.

Riemann's life was tragically cut short in 1866. He died of tuberculosis
at the age of thirty-nine in the village of Selasca on Lake Maggiore. He had
gone to Italy to attempt a cure. One of Riemann's greatest desires was to

unify the laws of electricity, magnetism, light, and gravity. Such a project was premature, but his mathematics still became the vital ingredient of the new physics to come. "Riemann left the real development of his ideas in the hands of some subsequent scientist whose genius as a physicist could rise to equal flights with his own as a mathematician," said mathematician Hermann Weyl. After a lapse of forty-nine years, that mission was at last fulfilled by Einstein. Had Riemann lived to a ripe old age, it is conceivable that Einstein might have thanked him in person.

The Maestro Enters

THE TALE HAS BEEN TOLD AND RETOLD so many times that it has taken on the strains of a fable. In some autobiographical notes Einstein remembered being haunted as a lad by a strange musing: If a man could keep pace with a beam of light, what would he see? Would he observe a wave of electromagnetic energy frozen in place like some glacial swell? "There seems to be no such thing," he recalled thinking at the youthful age of sixteen. Here was the seed for Einstein's casting out the absolute space and absolute time of Isaac Newton. Relativity arrived, not only from concerns over the flaws in Newton's mechanics, but also by contemplating the forces of electricity and magnetism as well as the mysteries of light.

For most of history it was generally assumed that light was something that was transmitted instantaneously. In a way it was everywhere "there." With this premise the light from a far-off star arrived at our eyes on Earth as soon as it was emitted. By the seventeenth century, however, certain thinkers had begun to wonder whether light had a finite speed after all—like sound— only far, far faster. In his great treatise on mechanics in 1638, Galileo may have been the first to suggest an experiment to test this hypothesis directly. One man stands on a hill and uncovers a lantern, signaling a companion positioned on another hill less than a mile away. The second man, as soon as he sees the initial beacon, flashes a return light of his own. Performing this task on hills spaced farther and farther apart, Galileo figured that the men would spot a successively longer delay between the dual flashes, which would reveal the speed of light. Members of the Florentine academy eventually carried out this experiment. Of course, they were unable to detect any delay, given the crudeness of Galileo's proposed method. Human reaction time is far too slow. The vast span of our solar system provided a far better test.

In the 1670s—Newton's day—the Danish mathematician and astronomer Ole Römer closely studied the movements of Jupiter's four largest moons, particularly the innermost one, Io. Specifically, he monitored the moment when Io periodically moves behind Jupiter and gets eclipsed. In doing this, he noticed that the interval between successive eclipses (an event that occurred about every forty-two hours) was not constant but regularly changed, depending on the position of Earth in relation to Jupiter. When Earth was moving away from Jupiter in its orbital motion around the Sun, the expected moment for Io to eclipse arrived later and later. This is because the light bringing that information to your eyes has to travel a bit more distance with each eclipse. By the time Earth reached its farthest point from Jupiter, Römer's measured delay had mounted up to twenty-two minutes (a better figure is sixteen and a half minutes). Others had noticed such changes before, but Römer shrewdly deduced that the delay was just the time needed for Io's light to traverse the extra width of Earth's orbit. Dividing Earth's orbital width of 186 million miles (299 million kilometers) by the delay time, Römer's crude measurements pegged a light speed of around 140,000 miles (225,000 kilometers) per second. That's quite fast, but Römer had demonstrated it was certainly not instantaneous. The modern value is 186,282 miles (299,792 kilometers) per second.

By the nineteenth century physicists had made great strides in understanding the nature of light. It was the era when scientists verified that light behaved like a wave. And, according to the physics of the 1800s, that required the light wave to propagate through some kind of medium. No reputable physicist would have dared to imagine that two objects could transmit light between each other without a substance to carry the wave. Sound waves move through air, and ocean waves travel through water; if there is a wave, something must be waving. For the heavens the transmitting agent came to be known as the "luminiferous ether," a reworking of the heavenly

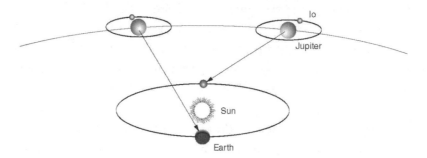

When Earth is farthest from Jupiter, the eclipse of Io is sighted later than
expected because the light must travel the extra width of Earth's orbit. In the
seventeenth century, Ole Römer used this effect to make the first good estimate
of the speed of light.

ether once postulated by the ancient Greeks as a godly essence that filled
the universe. As reconceived by the Victorian physicists, the ether was a
rather odd material: it had to be rigid enough to transmit light waves at ter-
rifically high speed yet still allow the Earth, planets, and stars to move
through it without resistance. Such a strange combination of properties
provided theoretical physicists with a cottage industry for more than a century. Scientific journals were filled with attempts to explain how the ether
could be both stiff and insubstantial. Permeating all of space, the ether also
served as a motionless reference system. In a way the ether resembled a
vast body of water. As a wave passes through the open sea, the water moves
only up and down. The wave is transmitting energy but not moving the
water forward. The same was thought to be true for the ether. Here at last
was Newton's absolute rest frame in physical form.

At the same time, explorations were taking place into the nature of electricity and magnetism. A link between these two phenomena first came in
1820 when the Danish physicist Hans Christian Ørsted discovered that an
electric current in a wire deflects a compass needle; in other words, a conducting wire acts as a magnet. Stop the current and the magnetism van-
ishes. The Englishman Michael Faraday completed the connection when he
noticed the opposite effect: a moving magnet creates an electrical current.
Born into poverty, with no formal mathematical training, Faraday was a
self-taught scientist. His mathematical deficiencies may have been to his

advantage. Highly visual, he imagined his magnetic objects being surrounded by fields of force, invisible lines influencing the movements of
objects within their midst. Such fields are wonderfully displayed in the
way iron filings align themselves when you sprinkle them around a magnet. Likewise, Faraday thought that electric fields somehow manipulated
charged particles with ghostly hands.

The distinguished Scotsman James Clerk Maxwell was just one in the
legion of people, both scientists and nonprofessionals alike, who were captivated by Faraday's experiments. A handsome man with a delicate consti-
tution, Maxwell became a professor of natural philosophy at the age of
twenty-four. Ten years later, with his august *Treatise on Electricity and Magnetism*, a triumph of nineteenth-century physics, he composed the mathematical words to explain Faraday's fields of force. His derivations, a set of
four partial differential equations so elegantly succinct that they show up
on physics students' T-shirts, demonstrate how electricity and magnetism,
two forces that on the surface seem so disparate, are merely two sides of the
same coin, each unable to exist in isolation. Maxwell united them into the
single force of electromagnetism.

More than that, Maxwell's equations also revealed that an oscillating electric current—charges moving rapidly back and forth—would generate
waves of electromagnetic energy coursing through space. He even worked
out the speed these waves would travel, which was related to the ratio of
certain electrical and magnetic properties. The result turned out to be exactly equal to the speed of light. Was this just a coincidence, as some
thought? Maxwell boldly said no. He concluded that light itself was a prop-
agating wave of electric and magnetic energy, an undulation that moved
outward in all directions from its source.

Visible light waves, each measuring about one fifty-thousandth of an
inch (one twenty-thousandth of a centimeter) from peak to peak, were just
a small selection of the wide range of waves possible. There could also be
waves of electromagnetic energy both shorter and longer than visible light.
The German physicist Heinrich Hertz proved this very fact in 1888. In a
laboratory humming with spark generators and oscillators, Hertz created
the first radio waves. Each wave had a length of thirty inches (around
seventy-six centimeters) and sped across his lab at the speed of light. It was

the first experimental verification of Maxwell's prediction that electromagnetic waves existed.

Maxwell died of abdominal cancer in 1879 at the age of forty-eight, nine years before Hertz conducted his experiments. In the year before his death Maxwell wondered about the problem that plagued physicists through most of that century: the motion of the Earth through the ether. Given the assumption that Earth was moving through a stationary ether, Maxwell thought of an optical experiment to detect an "ether wind" as Earth sailed at some sixty-seven thousand miles per hour in its annual voyage around the Sun. Like the air rushing past the passengers in a speeding convertible with its top down, the ether would be blowing past the Earth. Sparked by Maxwell's challenge, a young US naval officer, Albert A. Michelson, built a special instrument to spot the ethereal breeze in 1881 while assigned to Berlin for postgraduate work in physics. The instrument was his own design, an intricate assembly of mirrors and prisms that allowed pencil-thin beams of light to bounce back and forth in order to examine the wind. It came to be known as the Michelson interferometer, and with it Michelson found no hint of a draft. But Berlin traffic outside his laboratory rattled the detector at times, hurting its sensitivity. He tried again in 1887 as a civilian professor at the Case School of Applied Science in Cleveland, Ohio. There he teamed up with Edward Morley, a chemist at neighboring Western Reserve University. Using a vastly improved interferometer, the two researchers sent one beam of light "into the wind" in the direction of Earth's orbital motion and directed another beam at right angles to this path. Michelson once explained to his young daughter Dorothy how the test worked: "Two beams of light race against each other, like two swimmers, one struggling upstream and back, while the other, covering the same distance, just crosses and returns." The light beam fighting the ethereal "current" was expected to move a bit slower.

Michelson and Morley's elaborate equipment, set up in a basement laboratory, was mounted on a massive sandstone slab that floated on a pool of mercury to cut down on vibrations. But even with such precautions the two scientists detected no difference whatsoever in the measured velocity of their two beams of light, no matter which way the beams were pointed. The technique was so sensitive that it should have been able to measure a wind speed as small as one or two miles a second. With the Earth moving more

Albert A. Michelson in 1918.
Library of Congress.

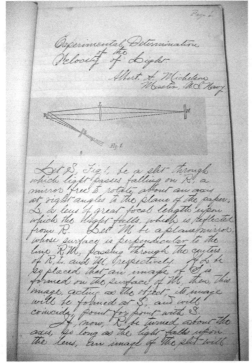

The first page of Albert
Michelson's handwritten draft
of a paper on the velocity of
light, written during his time
in the US Navy. *Case Western
Reserve University.*

than ten times faster, Michelson had figured that they would easily spot it. He was dismayed to discover that he was wrong. In 1907 Michelson became the first American to win a Nobel Prize in the sciences. He was honored for the development of his exquisitely sensitive optical instruments, many of them inspired by his futile search for the ether.

Others looked for the ether and failed as well. (Even Einstein, as a college student, wanted to build his own apparatus to measure the Earth's movement against the ether, but a skeptical teacher nixed the plan.) The null results forced physicists to come up with elaborate schemes to explain the lack of a detectable "wind." First the Irish physicist George FitzGerald and later Hendrik Lorentz, a Dutch physicist, suggested that an object traveling through the ether would contract—get physically compressed—in the direction of the motion. The dimensions of an object would somehow change as they moved about. That could explain why Michelson didn't notice a change; it was neatly canceled by this effect. Eventually, the great French mathematician Henri Poincaré railed against these complicated efforts to explain why physicists were not detecting Earth's absolute motion through the ether. In 1904 he presciently spoke of the need for a "principle of relativity." Soon such a theory was provided.

Historians continue to debate whether the Michelson-Morley experiment influenced Einstein in any way. He made only a brief and indirect reference to the test in his famous 1905 paper on relativity, even though his new theory neatly explained Michelson's continuing failure to discern the ether. What Einstein did stress was his perplexity over certain properties of electricity and magnetism. He was bothered by a seeming paradox. Consider either a bar magnet moving through a fixed coil or a coil moving over a stationary bar magnet. Each case is distinct. Maxwell's equations must handle each situation differently, depending on whether the coil is stationary and the magnet is moving or the magnet is stationary and the coil is moving. But each case leads to the exact same result: a current. Why should that be, asked Einstein? The descriptions of what is happening are different for the two distinct points of view, yet the observed outcome—the flow of an electric current in the coil—is the same. The experiment cannot reveal which object—the coil or magnet—is really moving in absolute space. Here was a crack in the notion of a fixed, eternal reference frame.

Newton's mechanics and Maxwell's equations of electromagnetism were each a monumental theory of their eras. Both yielded extremely accurate predictions. What disturbed Einstein was that these two great works of physics didn't seem to share the same rules on defining a space and time. Einstein's masterstroke was finding a way to make the two theories compatible with the simplest assumptions possible. Perhaps surprising was that his solution required no grand leaps of physics. Einstein's historic 1905 paper is exquisite in its simplicity. All of his hypotheses are based on the physics available in the nineteenth century. His one inventive assumption was a new conception of space and time. With that one change, all fell into place.

The prevailing image of Einstein has long been that of the venerable elder, the Chaplinesque figure with baggy sweater and fright-wig coiffure. But the youthful Einstein, at the height of his scientific prowess during the development of relativity, was a man whose limpid brown eyes, wavy hair, sensuous mouth, and violin playing aroused considerable attention, especially among women. One acquaintance compared Einstein's demeanor to that of a young Beethoven, full of life and laughter. Yet like the great romantic composer, the twentieth-century's most celebrated scientist had his dark side as well. He was also a loner (despite two marriages), a sharp-tongued cynic at times, and a self-centered man who could serve humanity yet express little empathy for the problems of those close to him.

Einstein was born in 1879, the year of Maxwell's death, into a nonreligious Jewish family, one well assimilated into the culture of southern Germany for more than two centuries. His father ran, with mixed success, an electrical engineering company, the high-tech business of its time. Early on Einstein expressed an intense desire to learn things in his own way. He apparently didn't talk until the age of three, stubbornly waiting until he could speak in complete sentences. Growing up with his younger sister Maja, little Albert loved doing puzzles, building structures, playing with magnets, and most especially solving geometry problems, the very key to his later work. He detested the German school system, with its emphasis on rote learning. And he didn't suffer fools gladly: he eventually dropped out of gymnasium (high school) due to conflicts with a teacher, among other reasons. Fortunately, he was able to enroll in a Swiss university, the Polytechnic

Albert Einstein in the early 1920s. *Library of Congress.*

in Zurich (the Federal Institute of Technology), although he was hardly a favored student among his professors. One called him a "lazy dog." As a result, he found no academic post upon graduation, surviving instead on the occasional teaching or tutoring assignment. Only in 1902 did he get a bona fide job, at the Swiss patent office in Bern. But all the while he was religiously devouring the books of the physics masters. He preferred self-learning. He had had that spark—that devotion—since childhood.

Physics was at a critical juncture at the turn of the twentieth century. X-rays, atoms, radioactivity, and electrons were just being discovered. It took a rebel—a cocky kid with mediocre college grades and no academic prospects but an unshakable faith in his own abilities—to blaze a trail through this new territory. Fearless at challenging the greats of his day, even as a student, Einstein was sure that the prevailing theory linking Newtonian mechanics with electromagnetism—electrodynamics—did not "correspond to reality . . . that it will be possible to present it in a simpler way."

To carry out this pursuit, his work as a patent examiner turned out to be a blessing. He fondly remembered his government office as "that secular cloister where I hatched my most beautiful ideas." Unencumbered by academic duties or pressures, Einstein was able to explore his ideas freely. By 1905, at the age of twenty-six, like a dormant plant that suddenly flowers, he burst forth with a series of papers published in the distinguished German journal *Annalen der Physik* (Annals of physics). Any one of his findings could have garnered a Nobel Prize. Inspired by the new quantum mechanics, he first proposed that light consists of discrete particles, what came to be known as photons (the Nobel winner). Second, he explained the jittery dance of microscopic particles—Brownian motion—as the buffets of surrounding atoms. It helped persuade the scientific community that atoms truly exist. Last, he submitted a paper blandly entitled "On the Electrodynamics of Moving Bodies" in which he revealed his special theory of relativity (actually rejected as a doctoral thesis topic for being too speculative).

Many of Einstein's mathematical arguments were similar to those already used by Lorentz and Poincaré, but there was a vital difference. Unlike his predecessors, Einstein was redefining the nature of time itself. He remarked many years later that the subject had been his "life for over seven years." Special relativity proposed that all the laws of physics (for both mechanics and electromagnetic processes) are the same for two frames of reference: one at rest and one moving at a constant velocity. It's as if Einstein were saying that a ball thrown up into the air on a train moving at a steady one hundred miles per hour on a straight track behaves just the same as a ball thrown upward from a motionless playground. For that to be true, though, means that the speed of light must also be the same in each environment, both on the train and on the ground. If the laws remain the same, each must measure the same speed of light. "[We will] introduce another postulate . . . ," wrote Einstein in his historic 1905 paper, "that light is always propagated in empty space with a definite velocity c, which is independent of the state of motion of the emitting body."

That appears reasonable, until you make the comparison fairly drastic. The effects of relativity are not really noticeable unless the comparative speeds are extremely high. So let's set up a test case: Consider a spaceship racing by the Earth at a steady 185,000 miles (298,000 kilometers) per second, just under the velocity of light by about a thousand miles per second. Common sense would lead you to believe that the astronauts are going nearly as fast as any light beam passing by, much as Einstein imagined as a youth. But that's not the case at all. The astronauts on that spaceship will still measure the velocity of a light beam at 186,282 miles per second, just as we do here on Earth. This situation seems bizarre, but not really. The speed of light remains constant, but other measurements get adjusted. The seeming paradoxes that arise are taken care of by acknowledging that time is not absolute. Time is, well, relative. The very term *velocity* (miles per hour or meters per second) involves keeping time, but the astronauts and Earthlings do not share the same time standard. That was Einstein's genius. He recognized that Newton's universal clock was a sham.

Since nothing can travel faster than the speed of light, two observers set apart in different frames of reference cannot really agree on what time it is. The finite speed of light prevents the two from synchronizing their watches. Einstein discovered that observers separated by distance and movement will not agree on when events in the universe are taking place. Consequently, by just looking, the Earthlings and astronauts will not agree on each other's measurements as well. Mass, length, and time are all adjustable, depending on one's individual frame of reference. Look from Earth at a clock on that spaceship that is whizzing past the planet. You will see time progressing more slowly than here on Earth. You will also see the spaceship foreshortened in the direction of its motion. Those on the spaceship, who perceive no changes in themselves or in their clock's progression, look at their home planet swiftly moving past them and see the same contraction and slowing of time in the Earthlings! Each of us measures a difference in the other to the same degree. Space shrinks and time slows down when two observers are uniformly speeding either toward or away from one another. Lorentz and FitzGerald spoke of an actual contraction in absolute space. Einstein, by contrast, showed that the changes are a perception of measurement. Space and time will be different in each reference frame. The only thing that the Earthlings and astronauts will agree on is the speed of light in a vacuum. It is the one universal constant. (Light does travel more slowly when transmitted through matter. Owing to the light's interactions with the matter along the way, its overall speed is effectively reduced.)

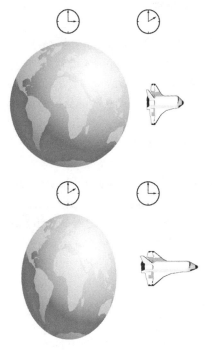

From Earth, a spaceship uniformly moving past the planet at near the speed of light will look shorter from end to end. Its clock will appear to tick more slowly (*top*). Those on the spaceship will perceive the same contraction and slowing of time on Earth (*bottom*).

With absolute time destroyed, there was also no need for absolute space. Our intuition that the solar system sits serenely at rest, with the spaceship speeding away in some motionless container of space, no longer works. It could just as easily be the astronauts at rest, with the Earth speeding away. The "introduction of a 'luminiferous ether' will prove to be superfluous," continued Einstein in his paper, "inasmuch as the view here to be developed will not require an 'absolutely stationary space' provided with special properties." Physicists no longer had to contend with awkward and complicated schemes involving a mysterious ether. There is no unique frame of reference that marks an absolute state of rest. Otherwise, any object moving in that fixed box would be able to catch up to a light wave. That explained why Michelson and Morley detected no ether wind. The fixed ether had been a fiction all along.

There is no need to dwell on speeding spaceships to perceive a relativistic effect. Relativity can be measured right here on Earth. Cosmic rays from space crashing into the upper atmosphere create muon particles—heavy

electrons—that spray downward at near the speed of light. Muons are extremely short-lived, lasting just millionths of a second, too little time for
them to reach Earth's surface. But experiments show that they do make it
to the ground. As relativity predicts, their inner clock appears to us to slow
down, which extends their life just long enough to make it to the surface.
From the muon's perspective, though, its lifetime is as short as it ever was;
it's the distance between the upper atmosphere and the ground that has
shortened, which allows the muon to reach the ground.

All is relative, even mass. As an object approaches the speed of light, its
mass increases noticeably, as measured by us. That's why nothing can go
faster than the speed of light. Its mass would be infinite at that stage; no
force exists that could push it faster, since the object would have infinite
resistance. Einstein later noted that light itself has an effective mass. And
because light is also energy, Einstein was able to link mass with energy in a
universal law. His calculations showed the relationship to be $E = mc^2$, where
c (as stated earlier) denotes the speed of light and m in this case is the rest
mass, the mass at zero velocity.

The teacher who once called Einstein a lazy dog, mathematician Hermann
Minkowski, brilliantly cut to the quick and discerned an even deeper beauty
in Einstein's new theory. ("I really wouldn't have thought Einstein capable of
that," he remarked to a colleague about Einstein's accomplishment.) With
his expert mathematical know-how, Minkowski recognized in 1907 that he
could recast special relativity into a geometric model. He showed that Einstein was essentially making time a fourth dimension. Space and time coalesce into an entity known as space-time. Time is the added dimension that
allows us to follow the entire history of an event. You can think of space-time
as a series of snapshots stacked together, tracing changes in space over the
seconds, minutes, and hours. Only now the snapshots are melded into an
unbreakable whole. Dimensionally, time is no different than space. "Hence-
forth," said Minkowski in a famous lecture the following year, "space by itself, and time by itself, are doomed to fade away into mere shadows, and only
a kind of union of the two will preserve an independent reality."

Six years earlier Minkowski had moved from Zurich to become a professor at Göttingen. Although he had made a number of important contributions to number theory and other areas of pure mathematics, he is largely

remembered for his reinterpretation of special relativity. It was easy for him to see that special relativity worked within a framework that had already been set up by mathematicians. "The physicists must now to some extent invent these concepts anew, laboriously carving a path for themselves across a jungle of obscurities, while very close by the mathematicians' highway, excellently laid out long ago, comfortably leads onwards," he said. To his mathematical eyes, special relativity was no more complicated than saying the world in space and time is a four-dimensional Riemannian manifold. To put it more plainly, Minkowski cleverly recognized that, whereas different observers in different situations may disagree on when and where an event occurred, they will agree on a combination of the two. From one position, an observer will measure a certain distance and time interval between two events. Perched in another frame of reference, a different observer may see more space or less time. But in both cases they will see that the *total* space-time separation is the same. The fundamental quantity becomes not space alone, or time alone, but rather a combination of all four dimensions at once—height, width, breadth, and time. Einstein, ever the physicist, was not impressed. When first acquainted with Minkowski's idea, he declared the abstract mathematical formulation "banal" and "a superfluous learnedness."

Often it is portrayed that the layman railed against special relativity when it was first introduced while the scientist greeted it with open arms. But for many scientists at this time, especially those deeply invested in classical physics of the nineteenth century, the concept of relativity was a psychological shock. Of course, opportunities to check the edicts of relativity when it was introduced were few and far between. It was only after several decades, when technologies had advanced, that measuring its effects became more commonplace. Some, though, would not accept special relativity on aesthetic grounds. William Magie, a professor of physics at Princeton University, stated in an address before the esteemed American Physical Society in 1911 that

> the abandonment of the hypothesis of an ether at the present time is a great and serious retrograde step in the development of speculative physics. . . . A description of phenomena in terms of four dimensions in space would be unsatisfactory to me as an explanation, because by no stretch of my imagination can I make myself believe in the reality of a fourth dimension. . . . A solution to be really serviceable must be intelligible to everybody, to the

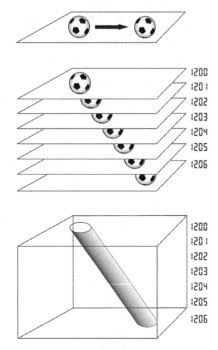

The simple movement of a soccer ball across a field is translated into space-time coordinates. With each tick of the clock, a snapshot captures the progression. Melded together, they form a tube that represents the entire motion in space-time.

common man as well as to the trained scholar. All previous physical theories have been thus intelligible. Can we venture to believe that the new space and time introduced by the principle of relativity are either thus intelligible now or will become so hereafter? A theory becomes intelligible when it is expressed in terms of the primary concepts of force, space and time, as they are understood by the whole race of man.

Critics were demanding that our ground-based experience, not mathematical formulas, be the criterion of truth. But they were shortsighted in believing that our earthly domain was the sole theater of experience. As the British astrophysicist Arthur Eddington noted in a lecture: "It has been left to Einstein to carry forward the revolution begun by Copernicus—to free our conception of nature from the terrestrial bias imported in it by the limitations of our earthbound experience." Before Copernicus, medieval scholars solemnly concluded that the Earth couldn't possibly be moving and turning. Otherwise, everything on the planet would be torn apart in the motion—clouds would get ripped out of the sky, and objects dropped toward a spinning Earth would obviously miss their mark because the Earth would have rotated around at great speed during the fall. Medieval thinkers had not yet mastered the concept of inertia, the tendency for objects to resist any change in their movement. (A falling object, already moving with the Earth's rotation, remains in sync as it drops.) When Copernicus placed the Sun at the cosmic hub, he thrust Earth into motion. He taught us to rethink our intuition based on new evidence. Einstein was doing the same.

Special relativity drew a line in the sand. On one side stood our past scientific history, when most physics theories could essentially be explained to the layperson. With a bit of hand waving and a reference to a mechanical model, a physical idea could be popularly illustrated. More important, the explanation did not violate the principles of common sense. But after 1905 the terrain suddenly changed. The world according to special relativity didn't seem to be describing our ordinary, humdrum surroundings. Simple mechanical models no longer worked.

There is a reason we are fooled: we live in a rather privileged place. Temperatures are extremely low (compared to a star, for instance), velocities are far from warp drive, and gravitational forces are essentially weak—an environment where the effects of relativity are very, very small. No wonder relativity appears strange to us. But as some physicists have put it, we are not free to adjust the nature of space-time to suit our prejudices. We're perfectly happy to adjust to the fact that thunder—a sound wave—arrives later than the lightning flash. It's part of our normal experience. Harder to accept is the finite and constant speed of light. Light travels so fast—it can wrap around the Earth nearly eight times in one second—that everything appears to occur simultaneously here on terra firma. It's difficult to experience directly the fact that observers, moving relative to one another, will disagree on the precise time an event occurred.

Now, special relativity was exactly that—special, a limited case. It dealt only with a specific type of motion: objects moving at a constant velocity. Einstein was determined to extend its rules to all types of motion, things that are speeding up, slowing down, or changing direction. But special relativity was "child's play," said Einstein, compared to the development of a general theory of relativity, one that would cover these other dynamical situations, in particular gravity. He tried to incorporate gravity directly into his special theory for a 1907 review article, but he came to recognize that it could not be done so readily.

Over the ensuing years Einstein's reputation would grow and soar. He finally left the Swiss patent office in 1909 when he received his first academic appointment at the University of Zurich. Two years later he moved on to the German University in Prague. After a year he went back to Zurich as a professor at his old haunt, the Polytechnic, where he had been so undistinguished as a student. He attained the peak of professional recognition when, in 1914, he moved to the University of Berlin as a full professor and member of the Prussian Academy of Sciences. Throughout these many years he waged a mental battle, amid teaching responsibilities, a failed marriage, and World War I. He struggled with the problem of recasting Newton's laws of gravity in light of relativity.

The first thing he recognized was that the forces we feel upon acceleration and the forces we feel when under the control of gravity are one and the same. In the jargon of physics, gravity and acceleration are "equivalent." There is no difference between being pulled down on the Earth by gravity or being pulled backward in an accelerating car. To arrive at this conclusion, Einstein imagined a windowless room far out in space, magically accelerated upward. Anyone in that room would find their feet pressed against the floor. In fact, without windows to serve as a check, you couldn't be sure you were in space. From the feel of your weight, you could as easily be standing quietly in a room on Earth. The Earth, with its gravitational field keeping you in place, and the magical space elevator are equivalent systems. Einstein reasoned that the fact that the laws of physics predict exactly the same behavior for objects in the accelerating room and in Earth's gravitational hold means that gravity and acceleration are, in some fashion, the same thing.

These thought experiments, which Einstein carried out liberally to get a handle on his questions, led to some interesting insights. Throw a ball outward in that accelerating elevator in space and the ball's path will appear to you to curve downward as the elevator moves upward. A light beam would behave in the same way. But because acceleration and gravity have identical effects, Einstein then realized that light should also be affected by gravity, being attracted (bent) when passing a massive gravitational body, such as the Sun. Driven by his powerful physical intuition, Einstein began to pursue these ideas more earnestly around 1911 while he was in Prague. At that time he was beginning to confirm that clocks would slow down in gravitational fields (an effect never before contemplated by physicists). He was also coming to understand that his final equations would likely be non-Euclidean. It was slowly dawning on him that gravity might involve curvatures of space-time. He was finally appreciating Minkowski's mathematical take on special relativity and its creation of space-time, that "banal" four-dimensional Riemannian manifold. Without Minkowski's earlier contribution, said Einstein contritely, the "general theory of relativity might have remained stuck in its diapers." Minkowski did not live to hear that; he had died in 1909 of appendicitis at the age of forty-four.

Returning to Zurich in August 1912, Einstein was eager to fashion his burgeoning conjectures into the proper mathematical format. Ignorant of non-Euclidean geometries, though, he joined up with mathematician Marcel Grossmann, an old college chum, to assist him in mastering the intricacies of this new mathematics. It was Grossmann who pointed out to Einstein that his ideas would best be expressed in the language of Riemann's geometry, by then advanced and extended by other geometers.

By the spring of 1913 their collaboration generated a paper with all the essential elements of a general theory of relativity. As science historian John Norton noted, "Einstein and Grossmann had come within a hair's breadth of . . . the final theory." But they backed off from their findings. The two convinced themselves, based on some misconceptions, that their equations could not reproduce Newton's laws of gravity for the simplest cases. Newton's laws might be incomplete, but they were not wrong. They would still hold when gravity was weak and velocities were low. But unable to retrieve Newton under those simpler conditions, Einstein and Grossmann abandoned this line of attack, assuming it was the wrong choice. This misunderstanding, as well as the knowledge that their equations were not as yet completely universal, kept them from grasping success. For the equations to work, they still had to use a special reference frame, which meant they hadn't met the standard of developing a general theory. In April 1914 Einstein moved from Zurich to Berlin, which ended the collaboration with his friend. At this point Einstein continued on his own, inexorably amending and tweaking his solutions, but now additionally armed with the mathematical insights introduced to him by Grossmann.

After working more than a year on the problem, Einstein became increasingly frustrated. His current theory, as it then stood in the autumn of 1915, could not accurately account for a particular motion in the orbit of Mercury. Einstein was then predicting a shift of eighteen arcseconds per century for this peculiar motion. He was aiming for the measured change of forty-five arcseconds (measurements today determine it to be forty-three). From his earliest days of contemplating a general theory of relativity, Einstein knew that a successful formulation of a new law of gravity would have to account for that anomaly.

The entire orbit of Mercury, a planet positioned about thirty-six million miles from the Sun, slowly revolves in the plane of the solar system. Imagine the orbit as an elongated ring. The point of the ring that is closest to the Sun—what is known as a planet's perihelion—shifts over time. For Mercury the perihelion advances about 574 arcseconds each century. A circular orbit encompasses 360 degrees. With 60 arcminutes in a degree and 60 arcseconds in each minute, 574 arcseconds is nearly 1/2,250 of an orbit. This means that Mercury's orbital axis makes a complete revolution roughly once every quarter million years. Most of this shift is due to Mercury's interaction with the other planets; their combined gravitational tugging alters the orbit. But that can account for only 531 arcseconds each century. The remaining 43 arcseconds were left unexplained, a nagging mystery to astronomers for decades. Newton's laws couldn't resolve the discrepancy, at least given the known makeup of the solar system. That led some to speculate that Venus might be heavier than previously thought or that Mercury had a tiny moon. The most popular solution suggested that another planet, dubbed "Vulcan" for the Roman god of fire, was orbiting closer to the Sun than Mercury, providing an extra gravitational pull. There were even a few reports of Vulcan sightings, but none were reliable.

Then Einstein noticed a mistake in one step of the derivations he had conducted with Grossmann. This spurred him to consider that approach once again. He began to modify the equations and in the process became aware of his earlier misunderstandings. This allowed him to begin seeing

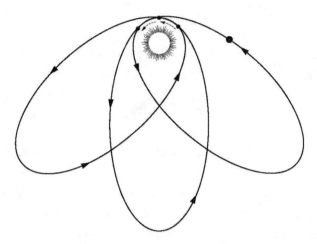

Over time, the point of Mercury's closest approach to the Sun—its perihelion—
advances. The perihelion makes a complete turn around the Sun in roughly
250,000 years. (The orbit's ellipticity is exaggerated for illustrative purposes.)

that he could recover Newton's equations when gravitational fields were
weak. His major effort took place during November 1915. On each of the
four Thursdays of that month he reported his incremental progress to the
Prussian Academy. A breakthrough came soon after his second report on
November 11. That week he was at last able to successfully calculate the
orbit of Mercury. He would later remark that he had palpitations of the
heart upon seeing this result: "I was beside myself with ecstasy for days." It
was the theory's first empirical success, grounding it in the real world.

Moreover, Einstein's new formulation also predicted that starlight would
get deflected around the Sun twice as much as he had earlier calculated
(and twice the amount if Newton's theory is used). That's because Newton's
laws take only space into account; Einstein now understood that gravity affects both space and time alike, which doubles the effect.

Triumph arrived on November 25, the day he presented his concluding
paper, entitled "The Field Equations of Gravitation." In this culminating
talk he presented the final modifications to his theory, which no longer
needed a special frame of reference. At last it was truly a *general* theory of
relativity. In a letter to fellow physicist Arnold Sommerfeld, Einstein noted
that he had just experienced "one of the most exciting, most strenuous
times of my life, also one of the most rewarding."

What he discovered by working within his new universal framework was the very origin of gravity. Written in the deceptively simple notation of tensor calculus, shorthand for a larger set of more complex equations, the general theory of relativity displays a mathematical elegance:

$$R_{\mu\nu} - \tfrac{1}{2}\, g_{\mu\nu}\, R = T_{\mu\nu}$$

On the left side of the equation are quantities that describe the gravitational field as a geometry of space-time. On the right side is a representation of mass-energy and how it is distributed. The equals sign sets up an intimate relationship between these two entities. The two are intertwined: matter becomes the generator of the geometry. Consequently, gravity is not a force in the usual sense. It is actually a response to the curvatures in space-time. Objects that appear to be manipulated by a force are just following the natural pathways along those curves. Light, as it gets bent, is following the twists and turns of the space-time highway. Mercury, being so close to the Sun, has more of a dip to contend with than planets farther out, which partly explains the extra shift in its orbit.

Space-time and mass-energy are the yin and the yang of the cosmos, each acting and reacting to the other. The very cause of gravity is rooted in this image: it is the manifestation of the geometry of space-time. What Riemann

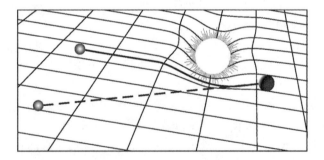

According to general relativity, space-time is like a vast rubber
sheet. Masses, such as the Sun, indent this flexible mat,
curving space-time. A star's light (solid line) traveling through
the cosmos follows these space-time curves. When our eyes
trace the light back as a straight-line path (dashed line), it
appears that the star has shifted its position in the celestial sky.

suspected, Einstein firmly established. Einstein was not at all influenced by Riemann's vague yet prescient thoughts about a metrical field (Riemann never imagined the necessary ingredient called space-time), but he was greatly beholden to Riemann's mathematics. Space, Einstein taught us, may be thought of not as an enormous empty expanse but as a sort of boundless rubber sheet. Such a sheet can be manipulated in many ways: it can be stretched or squeezed; it can be straightened or bent; it can even be indented in spots. This image of space-time as a two-dimensional sheet is often used to help us visualize the concept, but the curvatures, of course, are imprinted on the full four dimensions of space and time. So, massive stars like our Sun are actually sitting in a flexible, four-dimensional mat, creating deep depressions. Planets then circle the Sun, not because they are held by invisible lines of force, as Newton had us think, but because they are simply caught in the natural hollow carved out by the star. The more massive the object, the deeper the depression. Earth, for instance, is not holding on to an orbiting satellite with some phantom towline. Rather, the satellite is moving in a "straight" line—straight, that is, in its local frame of reference.

Think of two ancient explorers, who imagine the Earth as flat, walking directly north from the equator from separate locations. They move not one inch east or west but only push northward. Yet along the way they get word that they are moving closer to each other. They might then conclude that some mysterious force is pushing them together. A space traveler high above knows the truth. The Earth's surface is, of course, curved, and they are merely following the spherical contour. Likewise, a satellite is following the straightest route in the four-dimensional warp of space-time carved out by the Earth. As long as a heavenly body continues to exist, the indentations it creates in space-time will be part of the permanent landscape of the cosmos. What we think of as gravity—the tendency of two objects to be drawn toward each other—is a result of these indentations. Newton's empty box was suddenly gone. Space was no longer just an inert arena. Einstein showed us that space-time, the new physical quantity he introduced to physics, is a real-time *player* in the universe at large. Years later, reminiscing on this accomplishment, Einstein would write, "Newton, forgive me."

Starlight Waltz

IT WAS THE DISTINCTIVE ELEGANCE OF HIS theory that gave Einstein confidence in its validity. "Its artistry resides in its inevitability, the economy of its structure, the basic simplicity that shines through its complexities, and a pervasive beauty that, like all beauty, defies analysis," said Banesh Hoffmann, who once collaborated with Einstein. In 1930 Einstein himself wrote that he did "not consider the main significance of the general theory of relativity to be the prediction of some tiny observable effects, but rather the simplicity of its foundation and its consistency." And yet it would be one of those "tiny observable effects" that turned Einstein into a living legend.

Of course, the anomalous precession of Mercury was already known, and general relativity could explain it. But Einstein also predicted another effect, which Newton and others had once pondered but never pursued. As soon as physicists reliably thought of light as a wave in the nineteenth century, they generally assumed that it was different from matter, that it was immune from gravitational attractions. General relativity, on the other hand, asserted that light *must* be bent—that is, be attracted—by massive objects like the Sun, just like matter. More than that, the attraction would be twice the bending calculated when Newton's law of gravity is used. The extra contribution comes from the warping of space-time, especially near the mass, an effect that Newton's theory couldn't account for at all.

Observing how a star's light got deflected around the Sun was thus a way of detecting the curvatures—the sloping space-time valleys—that Einstein was proposing in his theory. Of course, it's not the light that is actually bending, although that is how it is commonly described. More correctly, it is light's space-time path being flexed. Near the Sun, a relatively lightweight star, the effect is very, very tiny. Einstein calculated that a ray of starlight just grazing the Sun's surface should get deflected by a mere 1.7 arcseconds (about one two-thousandth of a degree). That's roughly the width of the lead in a pencil seen from a football-field's length away. The bending diminishes the farther the light beam is situated away from the Sun and out of the solar space-time valley. Yet, despite being so minuscule, this effect was measurable with astronomical instruments.

In the spring of 1919, shortly after the close of World War I, Arthur Eddington, the British astronomer noted for his work on stellar physics, led a government-sponsored expedition to the tiny island of Principe, off the coast of West Africa, to look for this small light deflection during a solar eclipse. An eclipse offered the perfect opportunity to view a star near the Sun as the Moon blocked its dazzling glare. Luckily, the eclipse was occurring in a part of the celestial sky with an exceptional patch of bright stars. To minimize the risk of bad weather, another team of astronomers journeyed to the village of Sobral in northern Brazil to carry out the same measurement.

On the fateful day, May 29, Eddington and his teammate took sixteen photographs, most of them ultimately useless because of intervening clouds. "We have no time to snatch a glance at [the Sun]," wrote Eddington of his adventure. "We are conscious only of the weird half-light of the landscape and the hush of nature, broken by the calls of the observers, and beat of the metronome ticking out the 302 seconds of totality." In the end, two pictures turned out to have good images of the essential stars. Within a few days, as a precaution against any mishaps on the voyage back, one of the plates was examined on the spot. Eddington and his companion carefully compared the picture to another photo of the same celestial region, taken months earlier in England when the Sun was not in the way. Eddington, who freely admitted that he was unscientifically rooting for Einstein, was elated to see that the stars near the Sun had indeed shifted their apparent positions and by an amount that matched Einstein's prediction, give or take 20 to 30 percent. For Eddington that was close enough. It was certainly a larger bending than one would get using Newton's laws alone to calculate light's gravitational attraction. Here was proof, marginal as it was, that the long-reigning king of gravity, Newton, had been overthrown. Eddington later remarked that this was the most exciting moment in his life as an astronomer.

The Sobral expedition, which had fine weather and so was able to take many more photographs, confirmed the verdict. Einstein, ever confident, never doubted that the light deflection would be verified but was pleased, nonetheless, when he was informed via the scientific grapevine. He quickly dashed off a postcard to his mother to tell her the good news. The Royal Astronomical Society and the Royal Society of London, a scientific organization that Newton himself once presided over, held a special joint meeting that fall to officially announce the results, an example of the universality of science. A devastating war had just ended between Germany and Great Britain, and yet the British scientists were honoring a theoretical achievement made in enemy territory.

With its front-page reports of the solar eclipse experiment, the press on both sides of the Atlantic turned the name Einstein into a synonym for genius. His life in public was never the same again. Over the years celebrities, from presidents to movie stars, clamored to wine and dine him. He was besieged with autograph requests. Photographers and artists regularly arrived at his doorstep to do his portrait. Even Cole Porter included the acclaimed physicist's name in a 1943 song titled *It's Just Yours:* "Your charm is not that of Circe's with her swine / Your brain would never deflate the great Einstein." To this day his bushy mustache, helter-skelter hair, and world-weary eyes are instantly recognizable and remain an icon in cartoons and advertisements. "I have become rather like King Midas, except that everything turns not into gold but into a circus," remarked Einstein on his superstardom. For a man of thought, who yearned for a life of quiet contemplation, it was a state of affairs that he deemed "a dazzling misery."

Einstein died in 1955 and so did not live to see the further experimental triumphs of his theory in the latter half of the twentieth century. With the entrance of new astronomical techniques, light deflection experiments could be performed with much finer care than Einstein ever dreamed of. Solar eclipse experiments were carried out nine more times between 1922 and 1973, yet with only modest improvements in accuracy. Far better have been observations using a network of radio telescopes electronically linked around the globe. In this way one huge radio telescope as big as the Earth is created. Those who continued to question the validity of the coarse solar eclipse measurements were at last satisfied. By using this globe-spanning radio network to observe distant quasars, extremely intense and compact sources of radio waves, radio astronomers have been able to monitor how the apparent separation between close pairs of quasars changes as their radio signals pass close to the Sun. The accuracy in this type of measurement is nearly a thousand times better than Eddington's first crude try.

A more recent light deflection check was a space-age version of the 1919 test, but without the solar eclipse. The Hipparcos satellite, launched by the European Space Agency in 1989, spent four years making the most accurate measurements of stellar positions then assembled. It did this for stars down to a magnitude of ten (roughly fifteen hundred times fainter than the stars in the Big Dipper). The result: Einstein's prediction continues to hold up and with near perfection. In fact, Hipparcos's data were so precise that the Sun's ability to bend starlight could be detected halfway across the celestial sky. Stars located far from the Sun on the celestial sphere were observed to experience a shift in their apparent position, though far more weakly than for stars positioned closer to the Sun.

In 1964 Harvard-Smithsonian astronomer Irwin Shapiro, then with MIT's Lincoln Laboratory, came up with an entirely new and interesting variation to general relativity's light-bending effects. Shapiro suggested transmitting a radar pulse from Earth and reflecting it off another planet. This technique had already been used to measure distances to nearby planets. But if the pulse passed by the edge of the Sun, Shapiro figured that the radar signal's excursion to and from the planet should take a bit longer than what it would take were the Sun not there. That's because the Sun's warp of space-time, in a sense, adds a tad more distance to the journey; the radar beam must dip into the depression. Within two years the test was carried out. Radar signals were transmitted to both Venus and Mercury, as the planets were about to pass behind the Sun. For Venus the round-trip took about thirty minutes. Three hundred kilowatts of power were sent from a

radar transmitter at the Haystack observatory in northeastern Massachusetts; the echo that returned was as small as 10^{-21} watt. But that was enough
to notice that it took the signal about one five-thousandth of a second
longer to return to Earth when the signal passed near the Sun. The path
had lengthened by nearly forty miles. Later, signals from the Viking landers
on Mars, which set down in 1976, also were found to be delayed when
passing near the Sun. Agreement between the measurements and the
delay predicted by general relativity was within 0.1 percent, one part in a
thousand.

The most beautiful example of light bending in the universe is gravitational lensing. Take the case of Abell 2218, a compact and rich cluster of
galaxies situated more than a billion light-years from Earth. The view is
breathtaking. Several bulbous, elliptical galaxies sit like contented Buddhas
in the middle of Abell 2218. A number of bright disks—spiral galaxies most
likely—surround them. But there's more. Wispy arcs, 120 in all, encircle
the entire heart of the cluster. The streaks are arranged like the rings of a

This cluster of galaxies, Abell 2218, is so massive and dense that light rays arriving
from distant galaxies behind it are drastically bent to form arcs and rings. *Courtesy
of NASA/ESA/Andrew Fruchter and the ERO Team.*

dartboard. It is one of the universe's most wondrous illusions, created when Einsteinian light deflection is taken to the extreme.

When starlight passes by the Sun and gets bent, or deflected, the Sun is really acting like a lens. Recall that when you look through an optical lens it allows the object behind it to be magnified and brightened. It's a simple magnifying glass. A gravitational lens acts in a similar manner, only now it is gravity doing the work rather than a curved piece of glass. Soon after Eddington's successful solar eclipse test, Einstein and others discussed the possibility of light deflections—lensing—occurring farther out in space, as light passed by faraway stars. Depending on the orientation of the cosmic lens, objects behind it could be simply magnified or split into multiple images, as if some giant fun-house mirror were at work. But by 1936 Einstein concluded that "there is no great chance of observing this phenomenon" beyond the Sun, since the chances for two stars being properly aligned were too small.

Fortunately for astronomy, though, Fritz Zwicky at Caltech had a grander vision. In 1937 he declared that galaxies offered "a much better chance than stars for the observation of gravitational lens effects." Zwicky was right, although it would take four more decades before his visionary insight was confirmed. The first such cosmic lens was sighted in 1979 (totally by accident). Since then hundreds of lenses have been found. Some are single galaxies; others are entire clusters of galaxies, like Abell 2218. The cluster, trillions of times more massive than a single star like our Sun, collectively acts like a monstrous spyglass, greatly brightening the objects that lie far behind it. The faint blue arcs that surround Abell 2218 are actually the distorted ghostlike images of distant galaxies that reside some five to ten times farther out. This makes gravitational lensing more than a cosmic curiosity. As witnessed by Abell 2218, such lenses can act as a giant zoom lens. They take distant galaxies too faint to be seen and bring them into view. In this way astronomers manage a peek at the universe when it was far younger. No wonder that lensing has been called "Einstein's gift to astronomy." Awareness of lensing effects is actually becoming quite vital to astronomers. Otherwise, it can lead to some astronomical bloopers. When galaxy FSC 10214+4724 was discovered in 1991, for example, it was heralded as the most luminous galaxy in the universe. Though bright, it's not that brilliant. The Keck telescope in Hawaii later revealed that this galaxy is being brightened by a gravitational lens, a foreground galaxy located closer in. Oops, fooled by a gravitational illusion!

When Einstein proposed his theory of general relativity in 1915, he made another prediction that could not be detected as readily as light deflection. Scientists then had neither the instruments nor the techniques to measure this extremely tiny effect. It was Einstein's declaration that time will pass more slowly when nearer a source of gravitational attraction. To put it another way, a clock in space will tick more quickly than one weighed down by Earth's gravity. This situation is best imagined when we think of the gravitational force—the way that Einstein first did—as equivalent to the force one feels in an accelerating room out in space.

Picture a clock on the floor of that space elevator, with you on the ceiling observing it. But the room is accelerating upward. By the time the clock's ticks (marked by pulses of light) reach you at the ceiling, you and the ceiling have moved away in the motion upward. As the elevator moves faster and faster to mimic gravity, the peaks of the light waves will arrive at the ceiling at a slower rate. (In other words, the frequency will decrease.) Thus, the clock to you appears to be slowing down.

But, as Einstein taught us, the force experienced in this accelerating elevator is exactly mimicking the gravitational force on Earth. Hence, a clock on Earth would also tick away slower than one freely floating in space. This was a prediction unanticipated by any other physical theory. It was entirely new to physics. We don't notice this effect ourselves, for the atoms in our body are slowing down as well. We would know only by comparison. For example, any person who could miraculously survive on the surface of a neutron star, with a gravitational field a trillion times stronger than Earth's, would age noticeably slower than a person more loosely grounded on terra firma. While a decade passes by on Earth, Neutronians would experience around eight years. The movie *Interstellar* made dramatic use of this effect near the environment of a gargantuan black hole. Two astronauts landed on a planet that was closely orbiting the black hole. They stayed for only a few hours, but upon returning to their mother ship, they found that their left-behind shipmate was twenty-six years older. Black holes, the mightiest gravitational sinkholes in the universe, carry the time-dilation effect to the extreme. Should you approach the very edge of the hole itself, a fraction of a second ticks away for you, while an almost infinite number of eons pass by in the rest of the universe. Relativity, in this case, lives up to its name.

There's another way to look at this effect. You might think of light waves as springs bolted to the Earth—coils that get stretched as they ascend and attempt to climb out of the gravitational well dug into space-time by a massive celestial object. Shorter waves, such as blue or yellow light, would thus get longer as they soar upward, shifted toward the other end of the electromagnetic spectrum. They would get redder. Hence, the name for this effect—gravitational redshift. The reddening is so minuscule in the neighborhood of the Earth and Moon that scientists had to wait until 1959 to measure it. Robert Pound, along with his student Glen Rebka, detected the redshift by setting up an experiment on the campus of Harvard University. They measured how gamma rays shifted their frequency ever so slightly as the energetic waves either ascended or descended a seventy-four-foot-high tower within the Jefferson Physical Laboratory. The gamma rays came from a source of radioactive iron. To reduce the chance of the gamma rays getting absorbed by the dense air, a long Mylar bag was run through the tower and filled with lighter helium. The frequency changed within 10 percent of what Einstein predicted. Five years later, Pound and his colleague Joseph Snider got it down to 1 percent.

By the 1970s this gravitational redshift was being measured to levels of astounding accuracy. Atomic clock builder Robert Vessot of the Harvard-Smithsonian Center for Astrophysics rocketed into space one of his extremely accurate timepieces, a hydrogen maser clock, and compared its frequency to a similar clock on the ground. This special ninety-pound clock, so regular and exact that its time varied by about a billionth of a second each day (that's roughly equivalent to one second every three million years), was launched aboard a four-stage Scout D rocket from Wallops Island off the eastern shore of Virginia on June 18, 1976. The launch occurred near dawn. Impact was 118 minutes later in the mid-Atlantic Ocean, one thousand miles east of Bermuda. There was one nervous moment when Vessot and his colleagues lost contact with the spacecraft, but a minute later they reacquired the signal. A circuit breaker had accidentally cut off the power supply to the uplink transmitter. The basic idea of the experiment was simple—monitor the oscillations of the atomic clock as it traveled almost vertically up some six thousand miles and then down again. In the end they found that at six thousand miles above the Earth, where gravity has loosened its grip, the clock did indeed run a bit faster, by some four and a half parts in ten billion. If it had been there in orbit for seventy-three years, it would have gained a whole second compared to a clock on Earth. The accuracy of the test was within a hundredth of a percent, a hundred times better than the gravitational redshift measurements on the Harvard campus.

In 1976 such a test had little practical importance, but that's no longer true. The high-stability clocks aboard the Global Positioning System satellites, orbiting high above Earth, are regularly affected by the gravitational redshift. Twenty-four in all around the globe, these satellites must be synchronized to within fifty billionths of a second to allow users to know their position on the ground to less than thirteen feet (four meters). But without a relativistic correction, the clocks would run faster by forty-thousand billionths of a second each day, most of that due to the gravitational redshift. Periodic corrections are programmed in, otherwise the clocks would be out of sync within a minute and a half. Clifford Will knew that general relativity had finally arrived when he had to prepare a briefing on the theory for an air force general, as it became a matter of national security that GPS be as accurate as possible. Hollywood recognized the drama of this situation in the James Bond movie *Tomorrow Never Dies,* in which an evil genius attempts to insert errors into the system to send British ships into harm's way.

The gravitational redshift was not the only novel and strange effect predicted by general relativity. In a 1913 letter to the Austrian physicist and philosopher Ernst Mach, Einstein mentioned a new force that should come into play with general relativity. He called it "dragging." This was two years before Einstein had worked out his full theory. In many ways, dragging is to gravity what magnetism is to electricity. In fact, some call it gravitomagnetism. A charged particle as it spins creates a magnetic field that surrounds the particle; similarly, a spinning mass, such as the turning Earth, imparts a rotation to the surrounding medium, which is space-time itself! In 1918 two Austrian scientists, Josef Lense and Hans Thirring, calculated the effect such a spin would have. Consequently, it is sometimes called the Lense-Thirring effect.

As a celestial object spins, it twists space-time around itself.
NASA/Stanford/Gravity Probe B program.

Lense and Thirring saw that an object spinning pulls the very framework of space-time around with it, like a cake batter swirling around the beaters in an electric mixer. The whirling is strongest nearest the beaters and gradually diminishes farther away. In 1959 a magazine ad in *Physics Today* for a new kind of gyroscope sparked some physicists (while swimming naked in the Stanford University pool and musing as they exercised) to imagine the perfect gyroscope and how it could be used to measure this subtle feature of general relativity. By 1963 they obtained support from the National Aeronautics and Space Administration (NASA). For more than forty years, in fits and starts and rising like a phoenix to survive seven cancellations, the project moved forward. At a total cost of around $760 million, the endeavor was highly controversial. Called Gravity Probe B (Vessot's experiment was Gravity Probe A), it was one of the most expensive pure science projects that NASA had ever sponsored (and the longest in preparation).

In 2004 the project at last launched a set of four gyroscopes (essentially spinning wheels) four hundred miles (640 kilometers) above the Earth in polar orbit. For Gravity Probe B, the spinning gyroscopes were four quartz globes, each a mere one and a half inches in diameter, the size of Ping-Pong balls. The globes were coated with a layer of niobium, which gave them a silvery finish. They were even registered in the *Guinness Book of World Records* as the smoothest, roundest objects on Earth; they were polished to within forty atomic layers of being perfect spheres. Such perfection was needed to measure the tiny changes at work; distortions could have introduced a mechanical wobble that could have been mistaken for space dragging. Once set in motion and free of outside disturbances, the axes of these spinning globes kept pointing in the same direction. Because of conservation of angular momentum, the globes resisted any change in their orientation. This made them the perfect tool to measure space-time dragging. These gyroscopes spinning in space were aligned with a far-off star. But over time, as the spinning Earth dragged local space-time around itself, this alignment slowly drifted. Each gyroscope's axis, while maintaining its direction with respect to *local* space-time, was no longer aligned on the far-off star. This wobble was not a large effect. The image of the swirling batter is far too strong when it comes to Earth, a lightweight celestial object. Matching general relativity's prediction within 19 percent, the axis of each spaceborne gyroscope moved an infinitesimal 0.000011 of a degree each year because of Earth's dragging the framework of space-time around itself. That's the width of a human hair seen from a quarter mile away.

While this effect is virtually meaningless to Earth's cosmic life, such frame dragging has far bigger consequences in other environments, such as in quasars. A powerful young galaxy caught at the edge of the visible universe, a quasar emits the light of tens of normal galaxies, with most of that energy believed to be generated by a supermassive black hole in the quasar's center. The black hole contains the mass of hundreds of millions of suns. With such a large mass spinning, the magnitude of frame dragging is gargantuan. In fact, some speculate that it causes any nearby matter to spiral in toward the black hole's poles, which then shoots straight outward in spectacular jets that span hundreds of thousands of light-years. In this situation, frame dragging is made highly visible.

The noted Danish physicist Niels Bohr, who first conceived of an atom's inner structure, visited the United States in January 1939 to work a few months with Einstein at the Institute for Advanced Study in Princeton, New Jersey. But right before his ship, the MS *Drottningholm*, left Europe, he learned that nuclear fission had been discovered. German scientists had verified that their uranium nuclei were splitting into smaller pieces. Arriving at Princeton greatly excited by the news, Bohr immediately began working on the problem with John Archibald Wheeler, then twenty-seven, the Princeton physics department's newest addition and a specialist in atomic and nuclear physics. Together, they developed a general theory of nuclear fission. Using it they predicted that such nuclei as uranium-235 would be effective in sustaining chain fission reactions. Wheeler went on to become a central figure in subsequent historic developments in physics, including World War II's Manhattan Project to build the first atomic bomb and the later development of the hydrogen bomb.

Upon returning to Princeton in the early 1950s after this war work, though, Wheeler chose to move in a completely different direction. "I suppose it was infection," he said. "As a student I had read a book called *Problems of Modern Physics* by H. A. Lorentz, a great father figure in physics. And what were the problems? They were quantum physics and relativity." Having spent years on the first problem, Wheeler decided to tackle the second. It was a dicey decision. Relativity had turned into a backwater in physics, a field inhabited by lone specialists. "There were all these people working with Einstein who didn't know the rest of physics," recalled Wheeler.

For several decades general relativity had been the most admired yet least verified theory in physics. There was the subtle twist of a planetary orbit here, the tiny bending of a beam of starlight there. The theory could also account for the expanding universe verified by Edwin Hubble in the 1920s. Yet even with those successes, the experimental evidence was admittedly thin. Not until midcentury did things begin to change, largely due to the new technologies that could better assess the minute changes predicted by relativity. By the 1960s general relativists were entering a golden age of experimentation. Pound and Rebka at last measured the gravitational redshift, while Shapiro came up with an intriguing new way to measure space-time curvatures.

But this renaissance would not have occurred without another vital factor: a concerted effort by theorists in the United States, Europe, and the Soviet Union to study general relativity more deeply. And at the forefront of this movement to bring Einstein's theory back into the thick of mainstream physics and connect it to the universe at large was Wheeler. He played a large role in changing general relativity's moribund image, starting by immersing himself in the subject through teaching. "Much of the best teaching comes out of research, and much of the best research comes out of teaching," noted Wheeler. "If the class hour doesn't end with the teacher having learned something, he doesn't know how to teach." It was then that Wheeler encapsulated general relativity in one clear sentence: "Mass tells space-time how to curve, and space-time tells mass how to move."

Einstein gave the last seminar of his life in Wheeler's class, which met in the physics department's former quarters, the Palmer Physical Laboratory. It's an impressive gothic-style building, constructed of red brick with a thick slate roof. Erected in 1907, the building is now used as a center for Asian studies, so the physics themes played out in the stained glass windows and statuary of honored physicists of the past are oddly out of place. Wheeler slowly walked a visitor through his early years in physics. He proceeded past the heavy wooden doors at the entrance and up the wide central staircase. On the second floor, after a turn, the first room on the left is number 309. Here is where Einstein gave his last classroom lecture. The chairs, in dark wood, have widened arms on the right for note taking. Real blackboards, the old-fashioned kind, completely line the walls at the front and along the right side of the room. The seats are eight across and eight deep. The room has the smell of old wood and chalk dust, rather nostalgic and not unpleasant. One can almost picture the elderly Einstein at the front, in his casual attire, stepping the students through his thoughts. Wheeler recalled Einstein talking about three things: first, how he came to relativity; second, what relativity meant to him; and, third, why he didn't like quantum theory, whose edicts went against his scientific philosophy. The role of the observer is central to quantum theory; nothing is known until an observer measures it. Wheeler remembered Einstein wondering aloud, "If a being such as a mouse looks at the universe, does that change the state of the universe?" In such timeworn surroundings, Wheeler revived general relativity, taking it from its minor position in physics to one of its most thriving fields.

It began when Wheeler looked at a problem almost forgotten. What happens to a star that is particularly heavy? What happens to it at its death? J. Robert Oppenheimer (who later headed the Los Alamos Laboratory, the research arm of the Manhattan Project) and his student Hartland Snyder

John Archibald
Wheeler near Black
Hole, Nova Scotia, in
1981. *Roy Bishop,
Acadia University,
Courtesy of AIP Emilio
Segre Visual Archives.*

published a paper on this very question in the September 1, 1939, issue of *Physical Review.* (Coincidentally, Bohr and Wheeler had their paper on the theory of fission in the same issue.) They began with a star that has exhausted all its fuel. With the heat from its nuclear fires gone, the star's core becomes unable to support itself against the pull of its own gravity, and the stellar corpse begins to shrink. If this core is weightier than a certain mass, now believed to be two to three solar masses, Oppenheimer and Snyder confirmed that the core would neither turn into a white dwarf star (which will be our own Sun's fate) nor even settle down as a tiny ball of neutrons. From general relativity they calculated that the star would continue to contract indefinitely. It would be crushed to a "singularity," a condition of zero volume and infinite density conceived earlier by the German astronomer Karl Schwarzschild in 1915 when he was working with Einstein's newly published equations. For him, it was a place where all the current laws of physics break down. Oppenheimer and Snyder provided more details. The last light waves to flee before the door is irrevocably shut got so extended by the enormous pull of gravity (from visible, to infrared, to radio and beyond) that the rays became invisible and the star vanished from sight. What remained was a spherical region of space out of which nothing—no signal, not a glimmer of light or bit of matter—could escape. The ethereal boundary of this sphere has come to be known as the event horizon. It is not a solid surface but rather a gravitational point of no return. Once you've stepped inside that invisible border, there would be no way out, only a sure plummet into the singular abyss at the center. Space-time around the collapsed core becomes so warped that the stellar remnant literally closes itself off from the rest of the universe. "Only its gravitational field persists," wrote Oppenheimer and Snyder in their historic paper. For Schwarzschild such a condition arose as he worked out the first full solution to Einstein's equations. He had adopted a mathematical approach that allowed him to more easily map the gravitational field around a star. Having a singularity at the center was "clearly not physically meaningful," as he put it; Oppenheimer and Snyder, in contrast, were arguing that it could be the actual fate of a massive star.

But in 1939 Oppenheimer and Snyder didn't consider all of the forces that might possibly prevent such a dire finale. Coming back to the problem in the 1950s, Wheeler wondered if pressure, the resisting power of a substance, might change the result. Perhaps the pressure of the star's material would prevent the ultimate collapse. Or maybe in its death throes an aging star throws off so much radiation and matter that gravitational collapse is averted, and it settles down as a white dwarf or neutron star. "I was looking for a way out," said Wheeler. Kip Thorne, Wheeler's graduate student in the early 1960s, speculated that Wheeler's resistance to accepting the star's dark fate may also have been partly due to the idea originating with Oppenheimer. Wheeler, a political conservative, had reservations about Oppenheimer, who was long publicly challenged for his liberal beliefs. They had been on opposite sides during the first governmental debate on the need for a hydrogen bomb. "There was something about Oppenheimer's personality that did not appeal to me," confessed Wheeler in his autobiography. "He seemed to enjoy putting his own brilliance on display—showing off, to put it bluntly . . . I always felt that I had to have my guard up."

After Oppenheimer had worked briefly on the problem of "continued gravitational contraction," as he called it, he inexplicably dropped it, never taking it up again. "He didn't recognize the importance of it," explained Thorne. "But Oppenheimer's work with Snyder is, in retrospect, remarkably complete and an accurate mathematical description of the collapse of a black hole. It was hard for people of that era to understand the paper because the things that were being smoked out of the mathematics were so different from any mental picture of how things should behave in the universe." Wheeler was so skeptical, in fact, that he hardly mentioned the existence of Oppenheimer's paper in his early work on general relativity. His attitude didn't appreciably change until 1962 when David Beckedorff, an undergraduate student then completing a senior thesis at Princeton, reexamined the Oppenheimer solution and recast it in a simpler form. Beckedorff's new take allowed Wheeler and others to recognize that the collapse left behind a real object, even though its mass was hidden behind the event horizon. "It was a real eye-opener for me," said Thorne, just then starting his graduate work with Wheeler. Also crucial was the introduction of computers that could simulate the difficult physics of an imploding star. Wheeler was finally convinced that the star had to collapse. "Even if you put in the most powerful attempt to fight collapse, you can't prevent it," he said. "You always end up with a 'gravitationally completely collapsed object,'" as he was then awkwardly calling it.

When pulsars were discovered in 1967 and not yet understood, Wheeler repeatedly told the tale that a conference was quickly set up at NASA's Goddard Institute for Space Studies in New York City to discuss the possible suspects. Could they be red giant stars, white dwarfs, or neutron stars? According to Wheeler, he told the astronomers that they should consider the possibility that they were his gravitationally collapsed objects. "Well, after I used that phrase four or five times, somebody in the audience said, 'Why don't you call it a black hole?' So I adopted that," said Wheeler.

But the pulsar discovery was kept secret until February 1968, the announcement held off until the finding was published by the journal *Nature*. The pulsar conference did not take place until May. What is indisputable is that Wheeler used the phrase during an after-dinner talk at the annual meeting of the American Association for the Advancement of Science

(AAAS) in New York City on December 29, 1967. This lecture was later published in *American Scientist,* an article that is traditionally credited for the origin of the name *black hole.*

But the term was actually used much earlier, casually bandied about at the 1963 Texas Symposium for Relativistic Astrophysics. Albert Rosenfeld, then science editor for *Life* magazine, wrote an article on the Dallas meeting and mentioned that gravitational collapse might "result in an invisible 'black hole' in the universe." The phrase was mentioned a week later at an AAAS meeting held in Cleveland. There, *Science News Letter* reported, astronomers and physicists at the conference were suggesting that "space may be peppered with 'black holes.'" The physicist who used the term in Cleveland was Hong-Yee Chiu, then at the Goddard Institute. He had also attended the Texas Symposium. Today he claims that he borrowed the phrase from Princeton physicist Robert Dicke, both an experimentalist and a theorist on gravitation. In the early 1960s Chiu heard Dicke speak on the possibility that stars completely collapsed, creating an environment where gravity was so strong that no light or matter could escape. "To the astonished audience," recalled Chiu, "he jokingly added it was like the 'Black Hole of Calcutta,'" the infamous eighteenth-century prison where captives

Robert Dicke. *Courtesy of AIP Emilio Segre Visual Archives, Physics Today Collection.*

allegedly went in and never came out alive. Whoever originated the phrase, Wheeler still deserves much of the credit for its placement into the scientific lexicon. Wheeler's status bestowed a gravitas upon the name, giving the science community permission to embrace the term. It was, in fact, an appropriate physical description: a black hole is truly a pit dug into the fabric of space-time. "[Wheeler] simply started to use the name as though no other name had ever existed, as though everyone had already agreed that this was the right name," said Thorne.

While doing his graduate work at Princeton, Thorne observed firsthand this rebirth of general relativity. As experimentalists were testing their cherished theory in ways impossible to perform in the past, Thorne came to recognize that these scientists needed theorists like him to help them out. In general relativity it is not easy to decide what to measure. It's a slippery theory. Depending on the coordinates you use, you can come up with (what seem like) different answers. When carrying out his radar experiments, Shapiro was continually challenged, forced to justify his calculations to others again and again. It is one of the reasons that experimental relativity took nearly a century to bloom and flourish. It's a difficult business to decide what you will measure, the way in which you will measure it, and how to interpret the results. Many have stumbled along the way. Controversies have broken out over what is and what is not possible to observe. Partly to resolve these conflicts, theorists realized that they had to construct a more comprehensive scaffolding that contained not just Einstein's theory but alternate theories of gravity as well. "Although Einstein's theory of general relativity is conceptually a simple theory, it's computationally complex," said Thorne. "If you want to identify what it is you are testing in any given experiment, you basically need a larger framework than relativity itself gives. You need some other possibilities. You will then have a large class of conceivable theories of gravity in which relativity is one. These other theories then act as a foil against which to examine relativity."

Observers could set up experiments that tested for certain differences between these various theories, to see if Einstein's version held up. Clifford Will, who studied under Thorne at Caltech and is now at the University of Florida, and Kenneth Nordtvedt of Montana State University analyzed and categorized many of these alternative theories and even invented a few of their own. "Partly as strawmen," explained Will. "It was a way of motivating an experiment and interpreting the results." Others offered revised equations of gravity because they truly believed that general relativity needed to be amended for various theoretical reasons. All these theories, said Will, "forced general relativity to confront experiment as never before." The most famous alteration to Einstein's equations—and for a while its most serious challenger—was the Brans-Dicke theory.

Princeton University was the US center for the renaissance in general relativity and not solely in the theoretical arena. While Wheeler wrestled with gravitationally collapsed objects on paper, Dicke was reinvigorating relativity on the experimental end. "They thought about things in very different ways," noted Thorne. "Wheeler was a dreamer, driven to a great extent by physical intuition wrapped up in philosophy. Dicke was a gadgeteer who had theoretical dreams as well. But his ideas were radically different from Wheeler's. Wheeler thought about the universe in terms of geometry; Dicke thought about it in terms of field theory."

Dicke was a very generous scientist. In 1965 he helped Arno Penzias and Robert Wilson figure out that a pesky noise in their radio telescope at Bell Laboratories in New Jersey was actually the fossil whisper of the Big Bang, a buzz that had been echoing through the corridors of the universe for some fourteen billion years. Dicke was just getting ready to look for this cosmic microwave background himself. Penzias and Wilson later won the Nobel Prize in Physics for their accidental discovery, but Dicke was not bitter by this turn of events at all. He simply said he was "scooped."

Dicke, who died in 1997, was revered by an entire generation of physicists as the premier experimentalist of his time. "Dicke made experimental discoveries or elucidated theoretical principles that led to, among other things, the lock-in amplifier, the gas-cell atomic clock, the microwave radiometer, the laser, and the maser," Will has written. "He was involved, directly or indirectly, in so many discoveries for which Nobel Prizes were awarded, that many physicists regard it as a mystery (and some as a scandal) that he [never received] that honor." Trained in nuclear physics, Dicke began thinking about gravity around 1960 when he concluded that gravitational tests up to that point were woefully imprecise. One of his earliest ventures in this arena was to redo the famous experiments of Baron Roland von Eötvös of Hungary, who in 1889 and 1908 tested the equivalence of inertial mass (the aspect of a body that resists acceleration) and gravitational mass (the mass that feels a gravitational attraction) with exquisite precision. That mass is affected by these separate forces in the exact same way is the cornerstone of both Newtonian physics and general relativity. It is the reason that different masses, both light and heavy, fall at the same rate when dropped from a tall height (say, from the Leaning Tower of Pisa). A heavier mass is more attracted to the Earth than a lighter one. Yet at the same time it has a greater resistance to the acceleration—enough resistance to slow its progress and match the speed of its lighter companion in the overall fall. Eötvös found this match to be exact to a few parts in a billion. Dicke and his coworkers got it down to a few parts in one hundred billion, and later a Moscow team headed by Vladimir Braginsky made some improvements. In the mid-1990s a group led by Eric Adelberger at the University of Washington in Seattle reached the part-in-a-trillion level. They called their experiment "Eot-Wash," a pun on the good baron's name, which is pronounced "ut-vush."

In turning to these questions Dicke began thinking deeply about the foundations of gravitational theory itself. He came to believe in what is known as Mach's principle, named after Ernst Mach, who first voiced the concept decades earlier. Essentially, this principle states that the strength of gravity depends on the distribution of matter throughout the entire universe. If that is true, then gravity's forcefulness should diminish as the universe expands and diffuses its cosmic density. Dicke estimated at the time that the change would be roughly one part in twenty billion with each passing year. Einstein's general relativity did not allow for this at all. Working with his graduate student Carl Brans, Dicke incorporated Mach's principle into an alternate theory of gravity by adding an extra term to Einstein's equations. As a result, the Brans-Dicke theory came up with slightly different numbers on certain gravitational measurements, such as the perihelion of Mercury.

Einstein seemed to have gotten that right, but he had assumed that the Sun is fairly spherical. What if the Sun were more oblate, more squished down than people were aware of, perhaps due to a rapidly spinning core? If so, Einstein would be wrong and the Brans-Dicke theory more useful. To find out, Dicke set out to measure the Sun's oblateness to a finer degree than had ever before been done. Then a postdoctoral researcher (postdoc) at Princeton, Rainer Weiss recalled Dicke disappearing for a few weeks when he first got this idea. One Monday he walked back into the office with a fat sheaf of rolled-up drawings, about fifty or sixty. Dicke had stepped through the entire experiment in his mind. He had designed the telescope, the electronics, and all the optics, as well as the supporting structure. The two assistants assigned to build the apparatus eventually saw that Dicke had anticipated every correction, adjustments usually not discovered until an instrument is under construction. He had done it all with pure thought.

"We called it a very Dickesque experiment," said Kenneth Libbrecht, a former graduate student of Dicke's, "because all really subtle and clever experiments that had everything chopping back and forth were called that among my graduate student friends." Dicke's experiment involved gathering two differing signals in quick succession, which is the purpose of a lock-in amplifier. It automatically shifts a detection back and forth in synchronization. It will first take data of both a signal and its background and then measure just the background. The instrument is programmed to do this continuously, cycle upon cycle. In the end, by subtracting the background from the overall data, a weak signal can emerge from the noise. Dicke advanced the development of the lock-in amplifier and founded a company, Princeton Applied Research, to build such devices commercially. "As graduate students, we'd sit around and buzz about what Dicke was worth," recalled Libbrecht. "The rumor was about $10 million." His sole splurge was a cabin in Maine, where he spent a month each summer. "Other than that, he never acted rich. He wore his frumpy old tennis shoes to work, like everybody else."

Dicke, along with H. Mark Goldenberg, measured the Sun's roundness in 1966 by placing a circular disk—an occulting disk—in front of the Sun's image in the telescope. What remained was the very edge or limb of the Sun. Photodetectors basically scanned this thin circle of light to discern any fattening at the Sun's equator. Though the Sun does bulge because of its rotation, Dicke and Goldenberg reported far more bulging than previously assumed, enough to favor the Brans-Dicke theory over general relativity. It looked like Einstein was about to be overthrown. The sheer possibility is what motivated many to continue carrying out the classic tests of general relativity—light deflection, time delays, gravitational redshifts—to higher and higher degrees of precision so that the differing predictions of Einstein and Brans-Dicke could be appraised more rigorously. Over time, though, the enhanced measurements were found to match the predictions of Einstein's general theory of relativity far more than the predictions from Brans-Dicke or any other alternate theory of gravity. Yet Dicke's oblateness result, which seemed to hold up (or at least was not yet refuted), was still a lone and nagging concern. To settle the controversy, Libbrecht redid the solar measurements. Dicke had looked at the Sun from Princeton, which often has clouds in the sky. In the summer of 1983 Libbrecht set up a shack atop Mount Wilson just north of Pasadena in sunny California. "As a graduate student I was thinking, 'Not only am I going to knock down general relativity, but I'm going to revolutionize solar physics at the same time.' Then the whole thing fell down like a house of cards," said Libbrecht. All the effects that Dicke had analyzed over those many years appreciably disappeared. Princeton's cloudy skies, among other issues, had likely introduced errors. The Sun's oblateness is actually quite small. "That was Bob's last hurrah," noted Libbrecht. "But that's why he was such a hero. When the data said his theory was wrong, then for him his theory was wrong. It was as simple as that. He soon retired after that."

Pas de Deux

THROUGH THE WORK OF THEORISTS AND EXPERIMENTALISTS such as John Wheeler and Robert Dicke, space-time joined an ever-growing cast of characters in the universe's cosmic drama. This new participant took on a definite role and personality. Space-time became the universe's flexible stage, a rubbery structure that stars, planets, and galaxies could bend and dent in intriguing ways. This new outlook had dramatic consequences. It meant that, when an object embedded in space-time gets accelerated or jostled, it can generate ripples in this pliable space-time fabric. Jiggle a mass to and fro and it will send out waves of gravitational energy, akin to the way a ball that is bounced on a trampoline sends vibrations across the canvas. These gravitational (or gravity) waves will uniformly radiate outward much like light waves. But while electromagnetic waves move *through* space, gravity waves are undulations *in* space-time itself. They alternately stretch and squeeze space—stretch and squeeze somewhat like the bellows of an accordion in play. And as these ripples encounter planets, stars, and other cosmic objects, they will not be stopped. Rather they will simply pass right on through, as they expand and contract all the space around them.

Anything in the universe that has mass is capable of sending out gravity waves—all it has to do is accelerate. But the strength of the signal depends on the amount of mass and the nature of its movement. A mammoth body like a star has a powerful gravitational pull, but because it remains essentially stationary (aside from its steadfast motion within the galaxy), it emits little gravitational radiation. Earth also continually emits weak gravitational energy as it circles the Sun, although it would take the age of the universe before we'd notice any effects from the emission. The Moon sends out still weaker waves as it moves around the Earth. Even hopscotch players have an infinitesimal chance of emitting a gravity wave or two as they jump up and down.

The strongest waves, though, emanate from the most violent and abrupt motions the universe has to offer: stars crashing into one another, supernovas erupting, and black holes forming. Some of these events are not seen directly with electromagnetic radiation; thus, gravitational radiation offers a new means of exploring the universe. Gravity waves will not just extend our eyesight; they will provide an entirely new sense. "Gravitational waves could prove to be the most penetrating waves in nature. That is in part their charm but also their curse, since it makes them so difficult to detect," Rainer Weiss and former LIGO director Barry Barish have written.

Einstein first discussed the concept of gravitational radiation in 1916, shortly after he introduced his theory of general relativity. His paper on the topic was tucked away in the *Sitzungsberichte der Königlich Preussichen Akademie der Wissenschaften* (Proceedings of the Royal Prussian Academy of Science) next to articles on the perception of light by plants and the use of first person in Turkish grammar. An algebra mistake led Einstein to a misconception about the origin of gravity waves (what he called *gravitationswellen*) in this initial paper, but he made the correction in a follow-up paper in 1918. He recognized that just as electromagnetic waves, such as radio waves, are generated when electrical charges travel up and down an antenna, waves of gravitational radiation are produced when masses move about. Moreover, they would also travel at the speed of light. (As early as 1908, Henri Poincaré did mention that a relativistic theory of gravitation, yet to be fully established by Einstein, would likely involve the emission of gravitational waves, or *onde d'accélération*.)

To picture the generation of a gravity wave, Einstein imagined a cylindrical rod spinning around, like a game of spin the bottle. In this case the frequency of the gravity-wave emission would be twice that of the rotation. These waves would flow smoothly outward from the source, but because gravitational energy moving through space would disperse and grow weaker, which starlight does as well, Einstein doubted that gravity waves would ever be observed, even from the most violent astronomical sources. By the time the gravity waves from an exploding star in our galaxy strike Earth, for instance, they are little more than subatomic flutters. Were a gravity wave from a supernova in the center of the Milky Way to hit this page, it would be so weak that it would squeeze and stretch the sheet's dimensions by a mere hundred-thousandth of a trillionth of an inch—a measure ten thousand times smaller than the size of an atomic nucleus.

Given the extreme weakness of the signal, few scientists were interested in the phenomenon when Einstein first described it. Why bother with an effect too small to detect? Furthermore, a lively debate ensued for some four decades on whether gravity waves existed at all. Many seriously wondered whether they were just artifacts—unreal ghostly products—in the equations of relativity. That possibility inspired Arthur Eddington to ponder mischievously whether the waves really "traveled at the speed of thought." Even Einstein had doubts at one point while he was working at the Institute for Advanced Study. These suspicions lingered into the 1960s. The doubts initially arose because there is a pitfall in general relativity: its equations are written in such a way that they are independent of all coordinate systems. So when a theorist dives in and chooses a particular system of measurement, the results can be tricky to interpret. For example, if your measurement system happens to have its coordinates fastened to the masses, a gravity wave passing by wouldn't budge them off the coordinates (within your calculations, that is), leaving the impression that the wave had no effect on space-time whatsoever. "A few years ago here at Syracuse University," recalls gravitational-wave physicist Peter Saulson, "the department received a paper that seemed to prove that gravity waves would get absorbed by the interstellar medium. But it took the physicists here at the time several months to figure out exactly where the author of this paper went wrong. He had chosen a coordinate system unfamiliar to them. Each generation seems to have to work this out."

Today the controversy has been permanently laid to rest to everyone's satisfaction. Outside of a few cranks, no one questions anymore whether gravity waves truly exist (especially after the first one was detected in 2015). But even before that, with all the evidence that had been amassed in favor

Joseph Taylor. *Department of Physics, University of Illinois at Urbana-Champaign, Courtesy of AIP Emilio Segre Visual Archives.*

of Einstein's view of gravity, physicists were already convinced that gravitational radiation was a natural consequence of the theory. This early confidence was not based on faith alone. Indirect yet exquisite evidence that gravity waves were real had arrived in the 1970s, when two radio astronomers uncovered one of nature's most dependable gravity wave emitters in the celestial sky. Their tale of discovery is composed of one part ingenuity, one part serendipity, and two parts sheer pigheadedness.

Only a month after finishing his PhD in radio astronomy at Harvard University in 1967, Joseph Taylor heard about the discovery of a strange new object in the heavens. "This was a time when the journals were always publishing something quite new, but this was more unexpected than anything I can remember at the time," recalled Taylor. The discovery had been made using a sprawling radio telescope—more than two thousand dipole antennas lined up like rows of corn—near Cambridge University in Great Britain. Jocelyn Bell (now Burnell), then a Cambridge graduate student, was one of the laborers. "I like to say that I got my thesis with sledgehammering," she has joked. The telescope was designed by Cambridge radio astronomer Antony Hewish to search for quasars, and it was Bell's job to analyze its river of data. At a radio astronomy conference in 1983, Bell Burnell recalled the moment when she realized that some of the squiggles recorded on her reams of strip-chart paper didn't look quite right:

> We had a hundred feet of chart paper every day, seven days a week, and I operated it for six months, which meant that I was personally responsible for quite a few miles of chart recording.
>
> It was four hundred feet of chart paper before you got back to the same bit of sky, and I thought—having had all these marvelous lectures as a kid about the scientific method—that this was the ideal way to do science. With that quantity of data, no way are you going to remember what happened four hundred feet ago. You're going to come to each patch of sky absolutely fresh, and record it in a totally unbiased way. But actually, one underestimates the human brain. On a quarter inch of those four hundred feet, there was a little bit of what I call "scruff," which didn't look exactly like [man-made] interference and didn't look exactly like [quasar] scintillation. . . . After a while I began to remember that I had seen some of this unclassifiable scruff before, and what's more, I had seen it from the same patch of sky.

The 81.5-megahertz radio signal was emanating from a spot midway between the stars Vega and Altair. A higher-speed recording revealed that the signal was actually a precise succession of pulses spaced 1.3 seconds apart. The unprecedented clocklike beeps caused Hewish and his group to label the source LGM for "Little Green Men." This was done only half in jest. At one point some consideration was given to the possibility that the regular pulsations were coming from a beacon set up by an extraterrestrial civilization. Within a month Bell ferreted out, from the yards upon yards of strip chart that were spewing from the telescope, the telltale markings of a second suspicious source. Its period was 1.19 seconds. By the beginning of 1968, two more had been uncovered. When the phenomenon was announced to the public, a British journalist dubbed the freakish sources pulsars.

Remaining at Harvard as a postdoctoral fellow, Taylor quickly rounded up a team to observe the four pulsars with the imposing 300-foot-wide radio telescope at the National Radio Astronomy Observatory in Green Bank, West Virginia (a dish that dramatically collapsed in 1986). While the original pulsars had been found by searching visually for specific peaks—pulses—on a paper chart recorder, Taylor developed a different strategy to look for more. A pulsar, of course, beeps with a regular beat, but it also has a sort of echo. As the pulsar signal travels through the thin plasma of interstellar space, its various frequencies propagate at varying speeds: the high radio frequencies travel faster than the lower frequencies. (The speed of light can vary when not in a vacuum.) Consequently, like horses on a racetrack, the differing radio waves start spreading apart. They get dispersed. By the time they reach Earth, the high-frequency pulse arrives first, followed by the lower frequencies in quick succession. Overall, the pulse appears extended, sweeping rapidly downward in frequency. Taylor and his colleagues wrote a special program to look for this distinctive profile in their streams of celestial radio data, automating the search with the use of a computer. "No one had thought about doing this kind of thing before with a computer," says Taylor. With use of this strategy, the Harvard team found the fifth known pulsar. Within a year they found nearly half a dozen more, a sizable jump in number. And by then theorists at last figured out what a pulsar is.

Their models suggested that a pulsar is a neutron star, an object first imagined in the early 1930s. Soviet theorist Lev Landau had initially suggested that the compressed cores of massive stars might harbor "neutronic" matter. The neutron, a major constituent of the atom along with the proton and electron, was a hot item at the time, having been recently discovered by experimentalists. Caltech astronomers Walter Baade and Fritz Zwicky also picked up on the idea and proposed that under the most extreme conditions—during the explosion of a star to be exact—ordinary stars would transform themselves into naked spheres of neutrons. But their proposal was considered wildly speculative, and only a handful of physicists even bothered to ponder the construction of such a star. They would be so tiny that no one figured such minuscule stars would ever be detectable. Jocelyn Bell proved them wrong.

A neutron star squeezes the mass of our Sun into a space only a dozen or so miles in diameter. This occurs when the core of a particularly massive star runs out of fuel. No longer able to withstand the force of its own gravitation, the core collapses. The core had encompassed a volume about the size of the Earth, but in less than a second it becomes one giant atomic

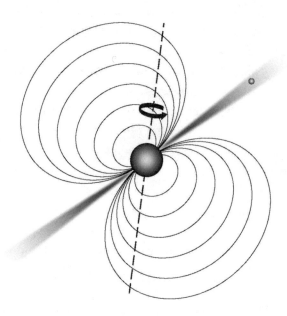

As a neutron star rapidly spins, radiation beams outward
from its magnetic poles. Sweeping around like a lighthouse
beacon, these beams are detected on Earth as clocklike pulses
(hence the term *pulsar*).

nucleus the size of Manhattan. All the positively charged protons and
negatively charged electrons are compressed to form a solid ball of neutron
particles. But like a compressed coil, the newly squeezed core rebounds a
bit, generating a powerful shock wave that, along with a flood of neutrinos,
eventually blows off the star's outer envelope as a brilliant supernova.

The neutron star that remains behind spins very fast. Like an ice skater
bringing in her arms—compressing herself to whirl ever faster—a collapsing
stellar core spins faster and faster during its compression as it conserves angular momentum. Its magnetic field grows very intense as well, to a trillion
gauss or so. (The Earth's magnetic field, by comparison, is a feeble half a
gauss, a hundred times weaker than the strength of a toy magnet.) Such a
rapidly spinning and highly magnetized body becomes no less than an electrical generator. As a result, narrow and intense beams of electromagnetic
waves are emitted from the neutron star's north and south magnetic poles. As
on Earth, these poles don't necessarily line up with the star's rotational axes.

So as the star spins around, the beams regularly sweep across earthbound telescopes, much the way a lighthouse beam skims across a coastline once every rotation. Radio telescopes sense the sweep as a periodic radio pulse.

In the fall of 1969 Taylor joined the faculty of the University of Massachusetts at Amherst to help establish the Five College Radio Astronomy Observatory in the woods of western Massachusetts. He viewed his continuing pulsar research as the perfect opportunity to study the final stages of stellar evolution. By then astronomers had detected a few dozen pulsars, but to truly understand them they needed a much larger sample. Researchers wanted to know how they were distributed through the Milky Way. Could all pulsars be associated with past supernova explosions? Just studying the radio pulses themselves seemed futile. "I have a friend in India who said that trying to understand a pulsar by looking at its radio pulses was like trying to understand the innards of a complicated factory by standing in the parking lot and listening to the squeaks of the machinery. In some cases, only a tenth of a percent of a pulsar's energy comes out in radio," noted Taylor. To get answers, Taylor wanted to double or triple the number of known pulsars by extending his computerized method for finding these radio beacons. He seemed destined for such a task since childhood.

Taylor grew up in the 1940s on a farm along the New Jersey shore of the Delaware River, just north of Philadelphia. Perhaps it was the farm machinery, but he and his older brother Hal became avid mechanics. They fiddled with all kinds of motors, both gasoline-driven and electrical. They even erected ham radio antennas on the roof of their family's three-story Victorian farmhouse. Taylor's interest in electronics continued through his undergraduate days at Haverford College, where he built a radio telescope for his senior thesis. He was able to detect the Sun and about five radio galaxies, some of the most distant objects then known in the universe. In Massachusetts, though, pulsars became Taylor's passion. He recognized that, with a large sample of pulsars on hand, astronomers could begin to use them as tools for probing interstellar space, seeing how the radio signals slowed, scattered, and polarized as they traversed the diffuse gases between the stars. Perhaps there were even new species of pulsars yet to be revealed. In his funding proposal to the National Science Foundation (NSF), Taylor did note—almost as an afterthought—that "even one example of a pulsar in a binary system [a pair of stars in orbit around one another] . . . could yield the pulsar mass, an extremely important number." It was a minor wish; he figured the odds would be against him. All the pulsars detected so far were solitary creatures. Since neutron stars are the remnant cores of exploded stars, it seemed reasonable to assume that the explosion would have disrupted the orbit of any companion star. Convinced of the merits of a large computerized pulsar search, though, the NSF allotted twenty thousand dollars for Taylor's project, a sizable sum in its day.

Needing assistance, Taylor sought out Russell Hulse, a graduate student then looking for a thesis project. Taylor offered him the perfect dissertation topic, a survey that would combine all three of Hulse's top interests: radio astronomy, physics, and computer science. Hulse readily accepted, titling his project, "A High Sensitivity Search for New Pulsars." Like Taylor, Hulse had been a tinkerer since his youth. When he was nine years old, he helped his father build a summer vacation home in upstate New York, putting in walls, rafters, and siding. "I was always building things," he said. "Fortunately, I came through the experience with all my fingers intact."

An eclectic child, he first went through his chemistry and biology phases, dissecting frogs and mixing chemicals. By the time he was thirteen, electronics had captured his fancy; that year Hulse was also admitted to the legendary Bronx High School of Science, notable for its Nobel Prize–winning graduates. Sparked by a library book on amateur radio astronomy, he built a radio telescope out of old television parts and army surplus in the backyard of his family's summer home. "Electronics was a lot more accessible than it is now," he recalled. "If you opened up a radio or TV set, there were all these parts: resistors, capacitors, tubes, wires, and coils. And you could take these electronic parts and build an antenna that had the potential to detect radio waves from the Milky Way."

For him it was magical, the idea of detecting signals out of the ether. "I don't have cable TV even today," he said. "I still get my signals the old-fashioned way, out of the air with an authentic antenna." His handmade telescope consisted of two flat sheets, each about four feet by eight feet, covered with wire mesh and meeting at right angles. Strung down the middle were a couple of dipole antennas. He tuned it to 180 megahertz, what would be channel 8 on a television dial. It didn't work, but he was never bored by the attempt. Through these experiences he developed a freewheel-ing, I'll-do-it-myself attitude that served him well in his schooling. He later taught himself programming on an early IBM computer at his undergradu-ate institution, Cooper Union, a college in lower Manhattan. One of his first programs was an orbital simulation.

Hulse chose the University of Massachusetts at Amherst for graduate work so he could combine his interest in electronics with astronomy. "Radio astronomy was still new, still rough-and-tumble," he said. And UMass, as it's familiarly called, was building a new radio telescope, what turned out to be an array of four 120-foot-wide radio dishes.

Hulse arrived on campus in 1970. By the time he was ready to tackle his thesis three years later, pulsars were still being found using a hodgepodge of techniques. His and Taylor's plan was to conduct a more systematic search taking advantage of the latest technology—called a minicomputer, although still as large as a couple of microwave ovens—to dig deeper into the galaxy for both weaker and faster pulsars. This required the use of the Arecibo Ob-servatory in Puerto Rico, the largest single radio telescope in the world. More than five decades in service, this telescope is legendary for the range of its observations. It made the first accurate measurement of Mercury's rotation, discovered planetary systems outside the solar system, and listened for signs of extraterrestrial life. Nestled in a natural bowl-shaped valley in central Puerto Rico, the telescope was initially built to explore a layer of the Earth's upper atmosphere known as the ionosphere. That required an antenna a thousand feet in diameter, encompassing the area of a dozen football fields. A limestone sinkhole in a valley near the town of Arecibo provided a natural framework for such a mammoth structure. The large collecting area also made it perfect for picking up pulsar signals, which are very weak.

The minicomputer, a Modcomp II/25, was programmed to sweep across a wide range of possible pulse periods and pulse widths in assembly-line fashion as the radio telescope scanned the sky over Puerto Rico. The aim was to look for a range of pulsars, ones that beeped as fast as thirty times a second or as slow as once every 3.3 seconds. It also looked over a range of dispersions, the amount of spread between a pulsar's high and low frequencies. All in all there were half a million possible combinations. "At each point in the sky scanned by the telescope," noted Hulse, "the

The Arecibo radio telescope, 1,000 feet (305 meters) in diameter, covers an area of about twenty acres. *Courtesy of the NAIC-Arecibo Observatory, a facility of the NSF.*

search algorithm examined these five hundred thousand combinations of dispersion, period, and pulse width." This made the search ten times more sensitive than previous surveys.

The computer was housed in two crude wooden boxes, a combination packing crate and equipment cabinet that Hulse had built out of plywood. It had thirty-two thousand bytes of core memory, a goodly amount for its day but millions of times less than today's smartphones, which can routinely incorporate tens of billions of bytes. A teletype was used for input and output, while a reel-to-reel tape drive stored the data. To get the maximum processing speed possible, Hulse programmed the computer in assembly code—the machine's internal digital language—using four thousand punch cards, an experience he does not long to repeat.

Hulse carted the minicomputer to Puerto Rico at a fortunate time. The telescope was then undergoing a major upgrade. Many observations were impossible at this time, but pulsar searching could still be carried on. This afforded him more time for his searches than he would have normally received, working in and around the construction and other people's observations. In fact, he stayed at Arecibo for some fourteen months—from December 1973 to January 1975—with only the occasional trip back to Massachusetts for a break.

The huge Arecibo dish does not move. It just serenely watches as the heavens continually revolve above it. To look for a potential pulsar as the telescope carried out its passive sweep, Hulse examined a particular spot on the sky for 136.5 seconds. Then he would begin to examine the next spot over. Hulse's prime-time viewing each day was when the plane of the Milky Way, toward the inner parts of our galaxy, passed overhead for some three hours.

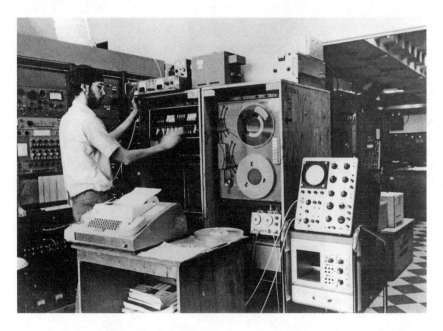

Russell Hulse shown in 1974 operating his computer and teletype at Arecibo observatory in Puerto Rico, where he searched for neutron stars. © *The Nobel Foundation.*

Just before that critical time, Hulse conducted the same routine. First he ran a tape, loading his program into the computer's memory. "As a classic computer hacker—and I'm using hacker in the positive sense—I had to make this program run fast enough so that all the data it collected over that three-hour observation window could be processed within twenty-four hours, before the next observation came up," he said. "I spoke fluent hexadecimal." During the observation itself, the computer would carry out the dispersion analysis and write its streams of digitized data onto a big magnetic tape. Over the remaining hours of the day, some twenty-one hours, the computer would review the resulting data and look for indicative signs of a pulsar. If the computer found a suspect, it would awaken the teletype and have it type out a cryptic line of information, which Hulse could easily translate. "You would know immediately," said Hulse. "The teletype started going *chunk-chunk-chunk-chunk*. It had a long line to print out if it found something. If there were interference, it was a nuisance, then you'd get all sorts of junk. Paper would start overflowing."

False leads could arrive from nearby thunderstorms, which were common during the summer in Puerto Rico. There was one pesky candidate signal that turned out to be emanating from an aircraft warning light on one of the telescope's support towers. And then there were the days that the US Navy held exercises off the coast. "I just sat in the control room," recalled Hulse, "watching signals from the naval radars . . . jump around on the observatory spectrum analyzer." But Hulse learned quickly to distinguish a false signal from a real one just by looking at the teletype print-out.

By the end of his fourteen-month stay, Hulse had cornered forty new pulsars, all located in the roughly 140 square degrees of the Milky Way observable with the big Arecibo dish. With each new find he drew a hash mark on the side of his trusty Modcomp II/25. All in all he had quintupled the number of known pulsars in that sector of the sky. That alone made a nice thesis for Hulse. But "it was of course eclipsed by the discovery of what was to become by far the most remarkable of these forty new pulsars, PSR 1913+16," he pointed out.

PSR is astronomical shorthand for pulsar, while 1913 is the pulsar's right ascension in celestial coordinates. It stands for nineteen hours and thirteen minutes. Astronomers have divided the sky into twenty-four-hour segments, akin to longitude but in this case the time it takes the Sun to make one complete circuit. The sixteen is the pulsar's declination or latitude on the celestial sphere. That placed the pulsar midway between the Aquila and Sagittarius constellations, close to the galactic plane that passes overhead at Arecibo. Hulse had started perusing this sector in the summer of 1974.

Things were pretty routine at that point. Hulse had already found about twenty-eight pulsars and by then even had pretyped forms to fill in the pertinent information on his finds. July 2 started out as just another day until one particular signal squeaked by the threshold that Hulse had set—just barely. With a bit of interference he would never have seen it. It was the teletype, automatically reporting any interesting finds, that first informed Hulse. The signal was an unusual candidate because it was particularly fast, with a period of about 58.98 milliseconds (17 beeps a second). "It would be the second-fastest pulsar known at that time, which made it exciting," said Hulse. (The faster pulsar was then the famous thirty-three-millisecond pulsar [thirty beeps a second] situated in the Crab Nebula, the remains of a supernova that was seen to explode in 1054.) Because the signal was so weak, though, Hulse was skeptical. "I put it on my suspect list," he said. "After the list got long enough, I'd devote a whole session reobserving them." Weeks later he was able to confirm the source. He proceeded to write down the signal's characteristics, tacking on a flourish at the end: "Fantastic!" he wrote on the bottom of his discovery sheet. His contact with Taylor was irregular, because phone service on the island was sporadic and e-mail was decades away. Through regular mail Hulse let his adviser know that he may have discovered a fast pulsar.

Hulse got back to PSR 1913+16 and his other suspects once again on August 25. This time it was the opportunity to measure their periods—the rates of their radio pulsing—more accurately. For the most part this was a standard and easy procedure: he just measured the candidate once and then measured it again about an hour or so later, to gather extra data for a more accurate measurement. But for PSR 1913+16 the period actually changed over that hour. The two measured periods differed by twenty-seven microseconds (0.000027 second). "An enormous amount," said Hulse, at least for a pulsar. "My reaction . . . was not 'Eureka—it's a discovery' but instead a rather annoyed 'Nuts—what's wrong now?'" Figuring

PSR *1913+16B*

RA	DEC	PERIOD	DM
19.13.74	16°00'	.059/0332	150
19 13 13	16°00'24"	/0018	167±5
±4s	±60"	59 98/24	
		,059030	

SCAN	DUMPS	DM CHNL	FREQ/tree	S/N
473	705-736	8,10	430/zB	7.25
526	1-128	8,9	430/28	13,0
535	1-384	8	430/28	15,0

$\ell = 49.95$ $b = 2.11$ fantastic!

Russell Hulse's discovery sheet showing his "fantastic" detection of PSR 1913+16, with its ever-changing periods scratched out by Hulse in frustration.
© *The Nobel Foundation.*

that it was an instrument error, he simply went back and measured it again another day. He kept marking down a new period on his discovery sheet, one after another. After the fourth one he scratched them all out in frustration. Obtaining accurate pulse periods was not a requirement for his thesis, but his compulsive nature took over. Perhaps his equipment wasn't sampling the pulsar fast enough to obtain an accurate fix on its period, thought Hulse. He then spent a full week writing a special computer program for the Arecibo mainframe to handle a faster data stream. He dropped all his other investigations and for two days solely observed this persnickety pulsar. But the problem only got worse. "Instead of a few data points that didn't make sense, I now had lots of data points that didn't make sense," said Hulse. Yet he did notice some regularities. He saw that the pulsing rate had decreased; the next day it decreased yet again. "My thesis wasn't going to fail if I didn't do this measurement, but beyond some point it became a sheer challenge," said Hulse. "I couldn't live with myself until I understood what was happening."

His thinking shifted at this point. He became convinced that the pulsar's period was actually changing, that it wasn't just an instrument error. He spent hours visualizing a spinning pulsar, trying to imagine how it might slow down. Finally, the image of a binary pulsar came to mind. Perhaps his undergraduate experience simulating stellar orbits paid off at last. At that point Hulse didn't know that Taylor had mentioned the possibility of finding such a system in the NSF proposal. In such a binary the pulsar would be orbiting another star. And that's why the pulsar's period varied. The pulsar period would regularly change—rise and fall, rise and fall—because of the orbital motion. When the pulsar moves toward the Earth, its pulses are piled closer together and its frequency appears to rise slightly; when it is moving away from us, the pulses get stretched and the frequency decreases. Optical astronomers have been acquainted with this effect for decades when observing the visible light of binary star systems. An audio version happens right here on Earth as well—the familiar rise and fall in the pitch of a train whistle as the train first races toward us and then away.

In his gut Hulse knew that he was right, but he had to see the turnaround, the moment when the pulsar started approaching the Earth in its orbit. If the pulsar were indeed a binary, its frequency should at some point start to increase. At last, on September 16, he saw it. His notebook records the proof. He had been processing the data in five-minute intervals, and every time the computer arrived at a period for that five-minute span, he marked it down on his graph paper. "I clearly remember chasing the period. Every one of those dots was a separate little triumph. The real exultation was seeing it hit the bottom and then turn around. There wasn't any doubt that it was a binary system. I drove back home that night, down the winding roads from the observatory, thinking 'Wow, I don't believe this is happening.'" It was also a relief. The chase had been stressful and had delayed work on his thesis.

Fairly soon, Hulse could tell that the pulsar was orbiting another object roughly once every eight hours. He quickly mailed Taylor a letter—a remarkably grumpy letter, actually—moaning about the extra work the pulsar had created for him. Today, Hulse says that the isolation and lack of sleep

As two neutron stars orbit each other, they generate gravitational waves in space-time that spread outward like ripples in a pond. *Courtesy of R. Hurt/Caltech-JPL.*

had probably gotten to him. But even with the letter on its way, he decided the news couldn't wait. With telephone connections so difficult from Arecibo, Hulse used the observatory's shortwave radio link to Cornell University. Cornell, in turn, patched the call via a phone line to Amherst. Taylor, immediately recognizing the import of Hulse's find, got someone to take over his classes and flew down to Puerto Rico within a couple of days with better pulsar-timing equipment.

Taylor and Hulse soon confirmed that the two objects were orbiting one another every seven hours and forty-five minutes. This meant that they were moving at a rather speedy clip, about two hundred miles per second, a thousandth the speed of light. While one was surely a neutron star, because of the pulsing, the other was likely a neutron star as well, because it was not big enough to eclipse the pulsar. (Pulsing is not detected in this second star because its beam is most likely not aimed at the Earth.) With such intriguing properties, Taylor and Hulse immediately recognized that they had been handed on a silver platter the perfect means for testing general relativity. Hulse remembers going to the library at Arecibo fairly soon after his find and consulting a copy of Misner, Thorne, and Wheeler's *Gravitation.* In their paper announcing the discovery, Taylor and Hulse wrote that the "binary configuration provides a nearly ideal relativity laboratory including an accurate clock in a high-speed, eccentric orbit and a strong gravitational field." Over just a few months they could actually detect the orbit of the binary system precessing, slowly dragging around. "That's the analog of the change in Mercury's orbit, but in this system it's much larger," said Taylor.

Up until the discovery of the binary pulsar, tests of general relativity were carried out primarily within our solar system. But with PSR 1913+16, the entire galaxy opened up to experimental testing of the rules of space-time. When Einstein derived the formula indicating that two objects orbiting each other would release gravity waves, he also recognized that the two objects would be drawn closer and closer together, owing to the loss of energy that the waves carry off into space. PSR 1913+16 was the perfect candidate to test this out. Here were two test masses—so compact and so dense—continually moving around each other. It was the ideal setup for detecting gravity waves (at least indirectly). Imagine a twirler's baton spinning in a pool of water. The motion would create a set of spiraling waves that move outward. Similarly, the motion of these two neutron stars should emit waves of gravitational energy that spread outward from the system. With energy leaving the binary system, the two neutron stars would then move closer together. At the same time their orbital period would get shorter. "As they depart for outer space, the gravitational waves push back on the [stars] in much the same way as a bullet kicks back on the gun that fires it," Thorne once explained. "The waves' push drives the [stars] closer together and up to higher speeds; that is, it makes them slowly spiral inward toward each other." But seeing such an effect required great patience. It could not be observed immediately but only over years.

While Hulse went on to other endeavors, Taylor and several colleagues, particularly Joel Weisberg, now at Carleton College in Minnesota, continued to travel to Arecibo to monitor the evolution of PSR 1913+16. (It was later renamed PSR B1913+16, to distinguish it as a binary pulsar.) Year by year they would spend two weeks—sometimes more—measuring the system as precisely as they could. Their major goal was to pinpoint the pulsar's timing. The tick of this pulsar clock is very regular, a sharp pulse every 0.059 second. Its blips are so regular and stable that its accuracy can rival the most accurate atomic clock on Earth. But to detect any changes in the binary's orbital motions, via the pulsar's precise tick, required extraordinary measurements. The system is located around twenty-one thousand light-years away, so its signal is very weak. Taylor's group had to build a special receiver that could better analyze the signal.

It took four years of monitoring before they at last detected a very slight change in the orbit of the two neutron stars. The answer arrived after analyzing some five million pulses. The orbit was definitely shrinking. The two stars were revolving around each other a little faster. This meant that the binary system was losing energy and the neutron stars were drawing closer together. Moreover, the energy loss was exactly what was expected if the system were losing energy in the form of gravity waves alone. It was a tough problem in relativity. Taylor first used an approximation that he found as a homework problem in a classic textbook on general relativity. But even before handling the calculation, he and his group had to make a number of corrections to the data. They had to correct for the motion of the Earth in the solar system as well as the perturbations introduced by the other planets. Variations in the Earth's rotation affect the signal's timing as well. There is also a slight delay of the signal due to the interstellar medium. "We measure and remove, measure and remove," said Taylor. They even had to adjust for the motion of the solar system around the center of the galaxy.

The gravity wave news was first released at the Ninth Texas Symposium on Relativistic Astrophysics, held in Munich, Germany, in December 1978. (The conference series originated in Texas, hence the name.) It was the highlight of the meeting. A report came out two months later in the journal *Nature*. Initially, though, there were doubts. Some wondered whether there was a third object in the system, which would upset the calculations. Or maybe dust and gas surrounded the pair, which could also explain the energy losses. But additional measurements over the years—with better and better receivers—only improved the accuracy. Taylor's graph, plotting the ever-decreasing orbital period, is a showpiece of science. The measured points lie smack-dab on the path laid down by general relativity. The measured energy loss owing to gravitational radiation agrees with theory to within a third of a percent. Such accuracy has been described as "a textbook example of science at its best."

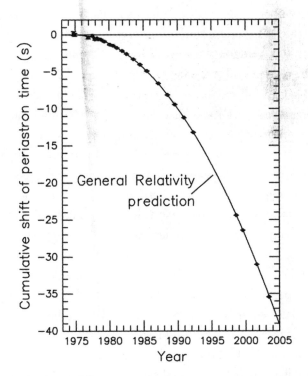

This graph shows the orbital decay of the binary pulsar PSR 1913+16 over the years. The data points, indicating the measurements, line up precisely on the line that illustrates the change predicted by general relativity. *Joel Weisberg and Joseph Taylor, in "Binary Radio Pulsars," ASP Conference Series, 2005.*

Each year the binary's orbital period decreases by about seventy-six millionths of a second. During each spin around each other, in the continuing pas de deux, the two neutron stars in PSR 1913+16 draw closer by a few millimeters. Over a year that adds up to 11.5 feet (3.5 meters). The two stars will collide in about three hundred million years. So clean and precise is this system that Taylor once remarked it was "as if we had designed the system ourselves and put it out there just to do this measurement."

In 1981 Taylor moved to Princeton University, but he continued to glean information on PSR 1913+16 from the simple ticks of its clock. After a couple of decades of measurement, some of the relativistic changes are fairly dramatic. The binary's orbital precession, for example, is quite vigorous. Taylor and his group now measure the change as 4.2 degrees per year;

that's thirty-five thousand times larger than the annual change in Mercury's orbit. The reason is clear: two neutron stars, so close together, affect the warping of space-time far more than our less dense Sun. Tightly bound together, they pack quite a wallop. Since its discovery, the binary pulsar has shifted its orbit a full half turn. Using this information, along with other orbital parameters, Taylor and his colleagues have been able to determine the masses of the two neutron stars to four decimal places. One is 1.4411 solar masses; the other is 1.3873. That's quite an accomplishment from a distance of twenty-one thousand light-years, given that each neutron star is a superdense nugget only a dozen or so miles wide. "One has to marvel at how much is learned from so sparse a signal," Rainer Weiss said.

The Hulse-Taylor binary, as it is now called, is no longer the sole member of its species. More than a hundred binary pulsars are now known, with most of the pulsars paired up with white dwarf stars rather than neutron stars. These particular pulsars spin much faster and so were not seen right away with older equipment. Stealing material from the white dwarf, the pulsar whirls itself up to faster and faster speeds, like a twirling ice skater on overdrive, hundreds of times each second.

Hulse left it to others to discover these new binaries. He actually departed the field of radio astronomy just a few years after his momentous find. He was not looking forward to the wandering life of a postdoc and the scarcity of secure academic positions. Wanting to be near his girlfriend, he first took a job at the Princeton Plasma Physics Laboratory in 1977, where he worked as a principal research scientist on computer modeling. In 2003 he moved to the University of Texas at Dallas. Hulse hasn't changed much since his graduate school days. He still has his beard and remains an ebullient and engaging talker, like a young kid explaining his favorite hobby. Taylor, meanwhile, continues his pulsar work, but now as a professor emeritus at Princeton. The team's old minicomputer is gone, long since cannibalized for parts, but Hulse did retain his original printouts, on newspaper-like green paper. "It's such hacker stuff. I read it now in a daze," he said with a chuckle. "I enjoyed doing it once."

Hulse had sighted the first pulsar of his extended search on December 8, 1973. Exactly twenty years later to the day, he was at a podium in Sweden delivering a lecture on the work he did to garner his doctorate. He and

Taylor had just received the 1993 Nobel Prize for Physics for their master-piece of measurement, one of the few times that prize over its century-long history has been awarded to astronomers. In the lecture Hulse described his work as "a story of intense preparation, long hours, serendipity, and a certain level of compulsive behavior that tries to make sense out of every-thing that one observes." He didn't ignore a troublesome observation. He tackled it with fervor, finding for Taylor and the astrophysics community the perfect laboratory for relativistic physics.

There had been controversy when Jocelyn Bell Burnell was denied a share of the 1974 Nobel Prize in Physics for her role in the discovery of the pulsar. The coveted award went to her adviser, Hewish, instead. This anti-student bias changed with the discovery of the binary pulsar. "It was very much a joint effort," said Taylor, a man well known in the astronomical community for his generosity and gentlemanly spirit, describing his work with Hulse. "Yes, one of us was a student, but there was no question that Hulse's work was an essential part of the operation." A longtime friend of Bell Burnell, Taylor invited Jocelyn to accompany him and his wife to Swe-den for the award ceremonies.

Bars and Measures

THROUGHOUT THE 1960S A CERTAIN QUESTION was regularly heard drifting through the hallways at general relativity conferences: "Has Joe Weber seen anything yet?"

Years before Joseph Taylor started obtaining indirect evidence for gravity waves, Joseph Weber was resolutely trying to catch one directly. It was a utopian crusade, and he knew it. At the time he started on his venture at the University of Maryland, his fellow physicists expected that it would require a full century of experimental work to attain such a goal. Even Weber admitted that "the probability of success under these circumstances had to be regarded as very small." But they admired his moxie.

Before Weber no one had even contemplated pursuing such an experiment. And there was good reason: "It had to scare any sane person," said Peter Saulson, for a gravity wave is such an incredibly tiny effect in our local surroundings. It was once calculated that if the great ocean liner *Titanic* were spinning once a second, it would generate less than a million billion billionth of a watt of gravity power. An atom bomb just ten yards from a detector would generate a gravity-wave signal more than a trillion, trillion times weaker than the one you could detect from a distant supernova exploding in our galaxy. It shows how difficult a task it is to create gravity waves that are detectable enough to run an experiment. Only events on a cosmic scale can provide the sources.

By convention the strength of a gravity wave is usually stated in terms of its "strain," a term borrowed from engineering. It's the fractional change in length—the magnitude of stretch—the wave would impart either in the distance between two masses or in a block of material. Two stellar-sized black holes merging in the center of our galaxy would certainly emit a substantial series of gravity waves. And even though gravity waves barely interact with matter, they'd still be deadly if you happened to be right there. Up close to the merging black holes, where the waves are forming, said Kip Thorne, "the nascent waves would stretch and squeeze an object that does not resist by about one-tenth of its size." This means that a six-foot man would be forcefully stretched and squeezed by more than half a foot hundreds of times over a single second. It's a frightening prospect. But fortunately, the energy of those waves spreads outward in all directions. Only a small portion reaches Earth after traveling tens of thousands of light-years, so by that time the strain is extremely small. Using meters to describe the effect, the strain would be a paltry 10^{-18} meter per meter. In other words, each meter along a rod would have its length changed by 10^{-18} (a millionth trillionth) of a meter, a span a thousand times smaller than the width of a proton. (The strain is the same no matter what unit of measurement you use; it's also 10^{-18} inch per inch and 10^{-18} foot per foot.) In any case, the cosmic tsunami is reduced to a quantum quiver. It's a cumulative effect. The longer the measured span, the greater the overall effect. It builds up over distance. How would such a strain, for example, affect the ninety-three-million-mile

Isaac Newton never imagined the existence of gravity waves, which alternately stretch and compress space-time and (to a lesser degree) any object as they pass by.

distance between the Earth and the Sun? With a strain on the order of 10^{-18} the gravity waves would expand and contract that space by a span about equal to the width of a germ. (For those unfamiliar with scientific notation, 10^{18} means 1,000,000,000,000,000,000, a one followed by eighteen zeros, an enormous number. Conversely, 10^{-18}, read as "ten to the minus eighteen," is the fraction 1/1,000,000,000,000,000,000, an extremely tiny number. 10^{-19} is a number ten times smaller than that, 10^{-20} is a hundred times smaller than 10^{-18}, and so on.)

Detecting such minuscule changes seemed an impossible task, but Weber fearlessly thought otherwise. He was partly impelled by his work at Maryland, teaching electrical engineering. "It seemed to me that if you could build an electromagnetic antenna to receive electromagnetic waves, you might be able to build a gravitational-wave antenna to receive gravitational waves," he said in recalling his thinking at the time. "I didn't know where they might come from. I just thought I'd start looking." In a way he wanted to be the Heinrich Hertz of gravity. Inspired by Maxwell's equations of electromagnetism, Hertz had confirmed Maxwell's prediction that electromagnetic waves existed. Weber was similarly determined to catch one of Einstein's predicted ripples in space-time. Astronomy was undergoing such a revolution after World War II, with the emergence of new disciplines such as radio and X-ray astronomy, that it was getting easier to contemplate finding the impossible. In the violent universe then being unveiled, thought Weber, nature might indeed be generating gravity waves capable of detection. Moreover, being the one to clinch one of the last predictions of general relativity was a powerful motivation.

Weber started seriously pursuing relativity research during the 1955–56 academic year while on sabbatical at the University of Leiden in the Netherlands, where John Wheeler was spending some time, and at the Institute for Advanced Study in Princeton, Einstein's old haunt. Both J. Robert Oppenheimer, then the institute's director, and Wheeler encouraged Weber's new interest. Later, his hopes would be nurtured by institute physicist Freeman Dyson, who calculated the gravitational waves emitted by the collapsing stellar core at the heart of a supernova. Dyson's results suggested at the time that the gravity-wave signal would be far stronger than previously expected.

In 1960, after a few years of investigating various approaches, Weber published his daring scheme for detecting gravitational radiation in *Physical Review*. In that and subsequent papers he outlined a clever technological trick for trapping a gravity wave. He surmised that a burst of gravitational energy moving through a solid cylinder would alternately squeeze and expand it ever so slightly, like an accordion in motion. The change would be incredibly small—as mentioned earlier, far less than the width of a nuclear particle. But then, long after the wave passed through, the bar would continue to ring. This phenomenon is similar to the vibrations that can be produced in a tuning fork when it is struck by sound waves. Similarly, a gravity wave that is tuned to the bar's natural acoustic frequency sets off a resonance, much like a gong continuing to ring after being struck. In both cases the dimensions and material of the bar or the gong determine which frequency of wave will trigger the ringing. Weber reasoned that he could position electronic devices on the sides of the cylinder and convert the extremely tiny gravity wave–induced movements into electrical signals that would then be recorded and scrutinized. This was not the first time Weber ventured into virgin scientific territory. Charting new landscapes was in his bones.

Weber was born in Paterson, New Jersey, in 1919 to Jewish immigrants and named Jonas Weber. His father was Lithuanian, his mother Latvian. The family name was originally Gerber, but that changed when his father, eager to move to the United States, took the visa of another man who had decided at the last minute to stay in Lithuania. With his family's prime language being Yiddish, Jonas mistakenly turned into Joseph when his mother first registered him for school.

Like many of his generation, Weber became fascinated by radios as an adolescent. He obtained his ham radio operator license at the age of eleven. Working as a golf caddie during the Depression for a dollar a day, he saved enough money for a book on electronics and started repairing radios after school. He was the youngest of four children, which he said was a blessing. While his older siblings were put to work early to help support the family, he was left alone to focus on his schoolwork. He decided to enter the US Naval Academy to break away from his immigrant roots and applied to a New Jersey senator, who relied on tests to make his appointment. Weber excelled and got the slot. Soon after graduating in 1940 with a degree in engineering, he was thrust into World War II. Assigned to the aircraft carrier *Lexington* as its radar officer, Weber shipped out of Pearl Harbor on December 5, 1941, two days before the Japanese attack. The following year he survived the carrier's sinking in the Battle of the Coral Sea. He finished out the war as commander of a submarine chaser in the Mediterranean. He also married his high school sweetheart, Anita Straus, with whom he had four sons.

After the war Weber extended his mastery of the new science of radar technology as head of the electronic countermeasures section of the Department of the Navy's Bureau of Ships. So great was his expertise that the University of Maryland hired him in 1948 at the age of twenty-nine as a professor of electrical engineering, with the stipulation that he also pursue his PhD. He did this at nearby Catholic University, where he obtained his degree in microwave spectroscopy in 1951. While completing his doctorate he worked out the concept of what later came to be known as a maser, the trailblazing forerunner of the laser. Instead of emitting visible light, a maser generates a beam of pure microwave energy—electromagnetic waves of kitchen-cooking fame intermediate in length between infrared and radar. Weber says he was inspired by a course in atomic physics. He was the first to publicly mention the maser principle at a meeting in Ottawa in 1952, the Electron Tube Research Conference, a yearly gathering devoted to the cutting edge of electronics. Weber never built his proposed device, though. For one, his calculations suggested that its performance would be minimal. Moreover, he had no research funds to construct a working model. Nobel Prizes went to those who did—the US physicist Charles Townes and two Russians, Nikolai Basov and Aleksandr Prokhorov—based on alternate mechanisms: "I was only a student in a way," said Weber. "I didn't know how the world worked."

Hearing visiting scholars lecture on general relativity at the university, Weber decided to use his 1955 sabbatical to study the subject in more depth. He did this, he said, because he "had wide interests and no money for quantum electronics." During this time he worked with Wheeler on the theory of gravitational radiation. It was a time when the debate was still raging on whether gravity waves truly existed or were just an artifact of the mathematics. That issue was finally settled at a seminal 1957 conference on gravitation held at the University of North Carolina in Chapel Hill, sparked by the mathematical insight of British theorist Felix Pirani, who beautifully demonstrated how gravitational waves were measurable. To put what he said simply, a gravity wave passing by would cause two test masses in space to move back and forth. Physicist Richard Feynman, also at the conference, extended this idea. He imagined two loose beads on a rod. When the gravity wave moves the beads, they would generate friction on the rod, heating it up. He and later physicist Hermann Bondi proved that gravity waves can deposit energy that is detectable (at least theoretically). These discussions at the conference may have been the moment that the idea to build a gravitational-wave detector was born, for Weber was in the audience. "My philosophy," Weber later recalled, "was to act like Galileo: build something, make it work, and see if you find anything."

Afterward, Weber spent nearly two years thinking of scheme after scheme for catching a gravity wave. He filled four three-hundred-page notebooks with possible detector designs. His 1960 *Physical Review* paper presented his most promising method for a gravity wave receiver: a gravity wave would hit an object—specifically, a piezoelectric crystal—whose vibrations would be converted into electrical signals and recorded. Piezoelectric crystals have the interesting property of generating a voltage when squeezed. Quartz is such a crystal. Pierre Curie first noticed this effect (and named it) in 1880. Finding it too costly to obtain a sizable bar of piezoelectric material, Weber and his lab associates came to realize they could bond the crystals to a much larger mass, a solid aluminum bar, which was cheap, easy to work with, and vibrated well. For a while Weber also wondered whether the entire Earth could be used as a detector. The Earth, if plucked like some musical instrument, would have various modes of oscillation, starting at one cycle every fifty-four minutes. Using a gravimeter, an instrument that measures the motion of the Earth's surface, Weber hoped to tune into the Earth's natural frequencies and see if it had been excited by a passing gravity wave. But background noises such as earthquakes, the motion of the oceans, and atmospheric events simply overwhelmed any potential signal. (Weber did persuade NASA to have Apollo 17 astronauts place a gravimeter on the Moon, where some thought there was a better chance of detecting a wave because of the Moon's lack of weather, quakes, or oceanic disturbances.) But in the end Weber concluded that his best bet was building a specialized detector in his laboratory.

Robert L. Forward, an employee of Hughes Aircraft Research Laboratories based in California, had arrived on the Maryland campus in 1958 to begin work toward a doctoral degree. Originally intending to do research on masers, Forward heard about Weber's pioneering venture and signed on. He obtained his PhD by constructing the first gravitational-wave antenna, along with David Zipoy. According to Forward, choosing its size was straightforward: "The reason for its particular length is that I said, 'Well, if I am going to have to manhandle that monster, let's make it this big.' I spread out my hands to what I could grasp, and we decided to make it five feet." Its width was two feet. The bar's total weight came to about 2,600 pounds. It was a fortunate choice of size. As soon as neutron stars and black holes moved from the realm of science fiction to real science over the 1960s, crude calculations (which largely still hold up) suggested that gravity waves emanating from these new celestial creatures would likely have frequencies of a few thousand hertz (a few thousand waves passing by each second), frequencies that could be picked up by solid cylinders of aluminum several feet long. (Before that the Maryland team had expected that its future bars would have to be hundreds of yards long to record the longer waves emanating from ordinary binary star systems.)

To isolate it from seismic disturbances, the bar was suspended on a fine steel wire in a vacuum tank that rested on acoustic filters to shield it from environmental disturbances. Surrounding the waist of the cylinder, like some bejeweled belt, were the piezoelectric crystals. According to Weber's strategy, once the bar was deformed by a wave—a gravitational burst—it would continue to oscillate and convert those vibrations into electrical signals. That's why these systems are also known as resonant bar antennas. The bars resonate like a bell in response to the passing gravitational wave.

Forward went back to Hughes in 1962, upon completing the bar's construction, but the bar continued to operate on the Maryland campus from 1963 to 1966. Pulses were recorded onto paper with a chart recorder, and the resulting pages scanned by eyeball. (Later, computers would be brought in for better, hands-off analysis.) But interpretation was difficult with one

Joseph Weber founded the field of gravitational wave astronomy in the 1960s with his invention of the bar detector. His reported detections were never confirmed. He is seen here in the early 1970s working on one of his bars at the University of Maryland. *Joseph Weber papers, Special Collections, University of Maryland Libraries.*

bar alone. The very motions of the atoms in the bar (on the quantum level, atoms are continually shaking) can swamp any gravity-induced signal. Weber believed that he could get around this problem by operating two bars simultaneously. He reasoned that the chances were low that the atoms would rattle around in each bar in the exact same way at the exact same time. Tremors occurring in both at once would then likely be due to an outside disturbance.

Additional bars were built and placed in a garagelike laboratory, what Weber called his gravitational-wave observatory, about a mile away. Small numbers of coincident pulses were observed. These results were reported in 1968, but interpretation was difficult. When a car accidentally ran into his outpost, generating a huge pulse, Weber realized that his detectors needed to be much farther apart, so that local disturbances could be ruled out as a possible source of interference. "Suppose you see a big pulse on one detector," noted Weber. "You can't be sure whether that pulse was due to a garbage truck colliding with the building, a lightning strike, or student unrest." (During the turbulent 1960s, student protesters painted obscene graffiti on his observatory.)

Weber's realization led to the construction of two identical bars, each five feet long, two feet wide, and weighing about 3,100 pounds. One was stationed on the Maryland campus, the other at the Argonne National Laboratory near Chicago, some seven hundred miles westward. The two detectors were linked by telephone line. If one of the detectors surpassed a certain threshold—what was determined to be the thermal background noise—the system was programmed to emit a pulse. If the other detector crossed that threshold at the same time, within 0.44 second, a coincidence marker was triggered. Most of the time the needle traces on their ever-moving chart paper displayed what was expected: random wiggles. But for a brief instant in December 1968, the needles on both detectors jumped at the same time. Over the next eighty-one days, Weber's team observed seventeen significant events shared by the two separated detectors. The researchers went through an exhaustive process to make sure the signals were real. They inserted time delays into their electronics to rule out random coincidences. If the signals were truly random, then inserting a time delay in one bar's circuits wouldn't have affected the coincidences at all. But the count fell, suggesting that something more than chance was at work. They made sure it wasn't some kind of electromagnetic disturbance, such as a solar flare or lightning. They monitored cosmic-ray showers and seismic disturbances at each site. In their estimation, nothing could explain the coincidences except the momentary passing of a gravitational wave. The cylinders' size determined the frequency: 1,660 hertz (cycles per second), in the range of frequencies expected to emanate from an exploding star.

It was ten years of work altogether, first arriving at a strategy and then setting up the instrumentation. Weber found it difficult to keep the discovery under his hat. He made his move at the Relativity Conference in the Midwest, a meeting held in Cincinnati in June 1969 and attended by

America's top relativists. Kip Thorne was there, reporting on the possible gravitational radiation generated by newly formed neutron stars. "And then Weber got up and announced that he had seen gravitational waves. It was quite a shocker to people," recalled the Caltech physicist. Thorne had known Weber for years and was highly interested in the techniques he was pioneering. "So I took it very seriously, as did nearly everyone else," he said. Conferees greeted the announcement with applause and tributes. Two weeks later Weber's official report was published in *Physical Review Letters,* the journal of choice for quick dissemination of a major discovery throughout the physics community.

Weber was now headline news. His picture was hard to miss, with his piercing eyes, resolute mouth, and a precision crewcut that had his hair standing at attention. The popular press wondered aloud in its stories whether Weber's discovery was the most important event in physics over the last half century. Weber's laboratory certainly became a magnet and an inspiration for physicists over the following months. Signing his visitor logbook in 1970 were William Fairbank of Stanford University and Ronald Drever from the University of Glasgow, who would soon establish their own gravity wave detection programs.

Weber was particularly interested in applying his find to astronomy. His announced detection came only two years after pulsars had been discovered, and he already had his eye on that phenomenon as a possible source for his gravity waves. By noting the period when his events were most prolific, Weber concluded that the gravity waves were arriving from the center of the Milky Way galaxy. His proclamation was eerily reminiscent of Karl Jansky's announcement in 1932 that radio waves emanated from the galactic center, an event that marks the birth of radio astronomy; at the time no one had ever expected such radio energies to come from the galactic core. According to the Weber team, the position of its two detectors offered a crude method for tracing the source of the alleged signal. At each site the long axis of each cylinder faced east-west. Situated this way, the strongest signals occurred when the source was straight overhead (or straight below the Earth, which poses no obstacle to a wave). Thus, the Sun could not be the source, they concluded; the signal didn't noticeably increase at noon or midnight, when the Sun was directly overhead or below. According to their initial statistics, the gravity-wave signals did peak when the center of our galaxy, located in the direction of the constellation Sagittarius, was in those prime positions. No one knew what events might be causing the blips on Weber's moving graph, but there were guesses: supernovas going off in the galaxy, colliding neutron stars, or matter falling into black holes, a term just coming into use.

Initial media coverage of Weber's announcement, even in the specialized physics press, was highly enthusiastic. It seemed that everyone was caught up in this new and exciting enterprise. It added a bit of élan to the field of general relativity. Here was a novel technique, never before attempted, to plumb the depths of the universe. Suddenly, any conference on relativity (normally a quiet affair) became a hot ticket; it was practically standing room only as people gathered to hear the latest news, exchange ideas, and compare notes. The response was not unlike the rush in 1989 to confirm the purported discovery of cold fusion. Within a year of Weber's discovery, at least ten groups were either proposing or already mounting similar searches. Work began in the Soviet Union, Scotland, Italy, Germany, Japan, and England. In the United States, gravitational-wave detector groups were eventually assembled at Bell Laboratories in New Jersey, IBM in New York, the University of Rochester also in New York, Louisiana State University (LSU), and Stanford University in California. Because of their fixed size, Weber's detectors were limited to recording one frequency, 1,660 hertz. The newcomers wanted to make detectors more sensitive and also tuned to additional frequencies in order to extend the science. At the time there were also a number of competing theories of general relativity avidly being discussed, and some of them, like the Brans-Dicke theory, predicted different effects when a gravity wave passed through a bar. So some hoped to use their new gravitational-wave detectors to test whether Einstein was right or wrong.

One of the first to put a detector together and check Weber's claim was Vladimir Braginsky, Dicke's counterpart in the Soviet Union. Stimulated by Weber's prediscovery publications on gravity-wave instrumentation, Braginsky began to write papers on both possible sources and methods of detection. Early on he recognized that thermal noise—the incessant jostling of a bar's very atoms—would produce major interference. By 1968 he was testing other types of bar materials, such as sapphire crystal, to see whether thermal noise could be reduced. Given this head start, he was able to construct a detector fairly quickly after Weber's pivotal announcement. Braginsky's group at Moscow State University later imagined building a series of bars of differing sizes—a sort of xylophone—to register different frequencies of waves at the same time. More than that, the Soviets began thinking of rocketing detectors into space. One scheme involved taking a detector shaped like a dumbbell far above Earth and rotating it. Theory suggested that a passing gravity wave would alter the rotation.

Elsewhere other imaginative schemes were devised. In Colorado researchers led by Judah Levine of the Joint Institute for Laboratory Astrophysics (JILA) ran tests in an abandoned mine, the Poor Man Relief mine located several miles west of Boulder in Four-Mile Canyon. There they reflected a laser light between two mirrors mounted on piers thirty yards apart. It was a crude setup. Supposedly, a standing wave could be set up between the mirrors, which a disturbance would alter. The changes would be noted by comparing the wave with a laser set to a constant frequency. R. Tucker Stebbins worked on the laser experiment while a graduate student at Colorado. "It used the Earth as the resonant bar, with a laser interferometer to detect the resonance," explained Stebbins. The instrument had already been set in place earlier to measure the speed of light more precisely, which required a quiet spot. The gravity-wave experiment ran a year or two but detected nothing except for some earthquakes and underground nuclear tests. The seismic interference was far too overwhelming.

Besides the solid cylinders of aluminum, which came to be called Weber bars, designers came up with new configurations, such as hollow squares, hoops, and U shapes. Others, such as Fairbank, world renowned for his work in low-temperature physics, began thinking about cooling the bar. To lower its temperature, the detector would be placed in a dewar, in effect a huge Thermos bottle. By lowering the bar's temperature to diminish atomic jitter, researchers could increase its sensitivity thousands of times over Weber's room-temperature system. Meanwhile, other investigators began to switch to a novel type of sensor to increase the bars' sensitivity. Instead of wrapping sensors around the cylinder, these designers placed a small diaphragm at the end of the bar. Physics decrees that any energy oscillating in the big mass will eventually transfer to the tiny mass on the end. But with that energy entering a smaller mass, the motion amplifies, making it easier to measure. While these new strategies were being developed, Weber worked on improving his own detectors.

Theorists were not immune to the excitement surrounding this new-found, experimental field. They were immediately drawn to figuring out what Weber was seeing, including British theorist Stephen Hawking. One of his earliest scientific papers analyzes the type of signal a gravitational-wave detector could register. He and his coauthor Gary Gibbons in 1971 argued that the bursts recorded by Weber could be coming from stars undergoing gravitational collapse to become neutron stars. But in working out the numbers, they cast an early dark cloud on Weber's work. They questioned the myriad events being reported. For the energies being detected, the newborn neutron stars had to be located within three hundred light-years of Earth. But by then the Maryland group was saying that it was seeing an event each day; there simply weren't that many neutron stars nearby.

What if the feeble signals were actually emanating from the Milky Way's center, around twenty-six thousand light-years distant, as Weber claimed? Then the energies had to grow tremendously. The energies had to be high enough to maintain a detectable strength by the time the waves reached Earth, situated in the suburbs of the galaxy. It was necessary for each pulse, radiating outward from the galactic core, to involve the conversion of roughly a sun's worth of mass into pure gravitational energy. But with Weber's events occurring daily, this meant that the galaxy would be losing mass at a tremendous rate, far too rapid to keep our galaxy intact from its birth more than ten billion years ago to the present day. The galaxy would be imploding at its center, long on its way to gravitational annihilation. If that was truly the source of Weber's signals, the galaxy should have been gobbled up by now. With such results, theorists started to become highly skeptical of Weber's assertions. "Either Joe Weber was wrong," commented one theorist, "or the whole universe is cockeyed."

By 1972 William Press and Kip Thorne wrote a review article of the field's accomplishments that posed the possibility that Weber might be wrong. But they did not yet reject Weber outright; they instead offered alternate possibilities: for one, the energy of the alleged gravity-wave source might be less than it appeared. Maybe an unknown source was nearer or the radiation somehow "beamed" in only one direction. Or possibly the gravitational radiation in the universe was far stronger than previously suspected. Could it be originating from outside the galaxy? "If these excitations are caused by gravitational radiation," wrote Press and Thorne, "then the characteristics of each burst are about what one expects from a 'strong' supernova or stellar collapse somewhere in our Galaxy; but the number of bursts observed is at least 1,000 times greater than current astrophysical ideas predict! Weber's observations lead one to consider the possibility that gravitational-wave as-tronomy will yield not just new data on known astrophysical phenomena (binary stars, pulsars, supernovas) but also entirely new phenomena (collid-ing black holes, cosmological gravitational waves, ???)." Their question marks left the door open to the unexpected.

By then, though, experimentalists were also beginning to experience grave doubts concerning Weber's discovery. Braginsky registered no sig-nals and simply shut his detector off. He felt that his energies would be better spent developing more sensitive detectors. David Douglass at Roch-ester and J. Anthony Tyson at Bell Laboratories also came up empty-handed. Tyson had been fascinated by the possibility of detecting gravity waves for nearly a decade, ever since he read a small monograph by Weber entitled *General Relativity and Gravitational Waves* as a graduate student at the Uni-versity of Chicago. Upon completion of a postdoctoral project in 1969, he was hired by Bell Labs to conduct research in low-temperature physics, his specialty, but Weber's historic announcement that year was too tempting to ignore. To corroborate Weber's detection, Tyson secretly set up two small bar detectors—each about three feet long and a foot wide—in his labora-tory, which was large enough for added equipment to get lost in the clutter. Though his bars were smaller than Weber's, he did use extremely low-noise amplifiers. Tyson ran his detectors covertly for about a year. "And I didn't see a thing," he said.

At last getting his employer's blessing to continue the investigation, Tyson built another bar that was appreciably larger than Weber's original detectors. It was roughly twelve feet long and two feet wide and weighed almost four tons. According to Tyson, it was capable of observing gravity-wave pulses that were far weaker than the events first recorded by Weber. But after operating steadily for a month in the summer of 1972, it regis-tered no special signal. Any surge that did appear was not beyond that expected from the random motions of the bar's atoms. Furthermore, Tyson arranged to observe the galactic center with one of the optical telescopes at the Cerro Tololo Observatory in Chile at the very same time that he was running his test. Nothing out of the ordinary was observed; no visual bursts matched up to the events Weber reported that he was seeing over that same period. Based on the signal Weber was allegedly receiving, said Tyson, "there should have been enough energy in other forms, such as electromagnetic waves, to knock your socks off. All you should have needed were binoculars."

Reporting his findings at the 1972 Texas Symposium held in New York City that year, Tyson got into a heated argument with Weber. Tyson had been supplied with four months of data taken earlier by Weber's group. The Bell Lab team tested for correlations with sunspots, temperature and barometric pressure differences near Weber's Maryland laboratory, and Earth strains midway between Argonne and Maryland because of lunar and solar tide effects. In the process Tyson and his colleagues discovered that with high probability the detected signals over that period aligned with changes in the Earth's magnetic field at the equator, an anomaly known as the disturbed storm-time factor. It is thought to be related to the ring current in the ionosphere around the equator. This was not proof that the Earth's magnetic field was the source of Weber's signals, but it opened up the question of whether some of Weber's events were of terrestrial origin.

Responding to Tyson, Weber stressed that his group did run magnetic tests on his bars, with field strengths far larger than the Earth's, and that his bars did not respond. He also countered that detecting a signal required at least two detectors because "coincidences are the result of externally produced signals, small compared with the noise of each detector!" You must compare to see anything, he said. At that same meeting Weber was upping the ante. He reported that his group was now seeing two or three coincidences per day.

Tyson conceded that he didn't as yet have a second detector for comparisons but stressed that his one detector was receiving nothing; there were no glitches of any kind above the normal noise levels. He was particularly distressed by Weber's lack of calibration. Weber had not yet tested his bar with a known source of energy to determine the exact strain it could see. Tyson himself used an electrostatic calibration. He applied a set force to the bar and noted the response. Amid this growing criticism, Weber continually offered one major rebuttal: all the other groups were simply not building the same instrumentation. He firmly believed that the sensors had to be placed around the waist of the bar, not on its side. "The scheme that works is not very difficult to set up," he declared. "[Until] you have reproduced the experiment that is known to work, I don't think I believe anything." He also contended that his competitors had inadequate temperature controls in their labs, burying any signals in excess noise.

Observers sympathetic to Weber's efforts sounded a similar refrain. Perhaps only Weber's bar, they noted, was actually tuned to the events erupting daily. Cardiff University sociologist Harry Collins, who has been conducting a sociological study of the field's development for more than four decades, even came across some overreaching supporters who briefly wondered whether psychokinesis was at work, with Weber as a focus.

The Maryland physicist did have more expertise. By then Weber had spent a dozen years designing, building, and testing his equipment before claiming his first signal. His competitors had spent, in some cases, less than a year. A colleague of Weber's told Collins that "one of the things that Weber gives his system, that none of the others do, is dedication—personal dedication—as an electrical engineer, which most of the other guys are not."

But eventually other bars identical to Weber's were constructed, and the news was not good for the founding father of the field. A collaboration in Europe—one group in Frascati, Italy, the other in Munich—built detectors that closely matched Weber's original design, including having the sensors wrapped around the bar's waist. One of their tests ran for 150 days, intermittently from July 1973 to May 1974. Expecting to see at least one pulse a day, like Weber, they ended up seeing nothing at all. The barrage of these negative reports eventually reached a critical mass, erupting in a confrontation that has become legendary in the history of the field.

Richard Garwin, a maverick physicist at IBM known for his scientific crusades and a gravitational-wave novice, decided to build an antenna to settle the issue once and for all. Garwin, who in his early twenties had helped design the first hydrogen bomb, was very suspicious of Weber's statements and wary of his statistics. With a colleague at IBM, James

Levine, he built a 260-pound detector within six months. Running a month in 1973, the small aluminum bar picked up one pulse, likely a noise. Garwin later learned from David Douglass that at least some of Weber's daily signals may have been the result of a computer mistake. Douglass had noticed a programming error that would register a coincidence between Weber's two antennas when none actually occurred. Nearly all the "real" coincidences reported by Weber over one five-day period could be traced to this error. He was seeing a signal in what most likely was pure noise. Weber was also claiming to see coincidences between his detectors and an independent detector run by Douglass in Rochester. But that couldn't have been possible. The two labs, it was discovered, used different time standards. One lab used Eastern Standard Time, the other Greenwich Mean Time. The data that Weber had been comparing (and which appeared to show coincidences) were in reality four hours apart. For Garwin and others this seemed to prove that Weber was selectively biasing his data to suit his claims.

In a surprise attack, Garwin verbally confronted Weber with this information at the Fifth Cambridge Conference on Relativity held at MIT in June 1974. An acrimonious exchange ensued at the front of the lecture hall. As the two approached each other with clenched fists, moderator Philip Morrison, disabled by childhood polio, raised his cane to keep them apart until the tension subsided. Their battle continued, though, in an exchange of terse letters in *Physics Today,* the magazine of record for the American Institute of Physics. Garwin contended that the very way in which Weber defined a coincidence introduced errors. Analyzed in a different manner, the purported signal went away. Garwin and his associates carried out a computer simulation to demonstrate this effect. Though their data were random, they could coax what looked like a signal out of sheer noise.

For many, this confrontation reminded them of an earlier contretemps in 1970, when Weber first claimed that his signals were emanating from the galactic center. At a lecture, he announced that the peak reception occurred every twenty-four hours, at a time when the Milky Way's center was overhead. Then someone in the audience pointed out that Weber's reception should also be good when the galactic center was directly below his antennas (the Earth being no barrier to a gravity wave). It wasn't long before Weber was reporting that his peaks were indeed arriving every twelve hours. (To boost the event rate for a better statistical analysis, Weber contended his group had been "folding" the second twelve hours of data over the first twelve hours since those scans above and below the Earth match each other. This, he says, resulted in his initial misstatement.)

Though not backing down from his conclusions, Weber did initiate a number of checks and balances at his laboratory both before and after this confrontation with his critics. First of all, he removed himself from direct participation in the data reduction to eliminate any personal bias. He put it into the hands of graduate students, postdocs, and associates. Any data recorded on pen-and-ink charts or signals picked out by eye were now cross-checked with automatic computer systems. Artificial pulses were inserted and identified by programmers, unaware of the timing. Telephone circuits connecting Maryland and Argonne were checked for spurious noises. From Weber's point of view, he answered all criticisms and corrected all possible sources of error, but the damage was done. From that point on, his announcements and papers were viewed with growing distrust by others in the field. They were leery of his scientific methods and frustrated that his publications at times were hazy on details.

Just two weeks after the infamous 1974 Cambridge meeting, the Seventh International Conference on General Relativity and Gravitation was held in Tel Aviv. The final session gathered four key participants in the bar experiments: Joe Weber, Tony Tyson, Ron Drever from Glasgow, and Peter Kafka from Munich. Kafka reported on the negative Munich-Frascati results, while Tyson talked about his latest failed effort. In collaboration with colleagues from the University of Rochester, he had constructed an even larger bar and began operating it in 1973. This bar eventually ran, off and on, for some eight years. Partway through that period, its twin was built by Douglass at Rochester, and the two detectors ran in coincidence for some 440 days from 1979 to 1981. Again, not one cosmic shiver was registered. Tyson jokingly called his last gravity-wave antenna "the most expensive thermometer in the world."

During the Tel Aviv panel discussion, Drever reported on his ongoing work in Glasgow. His device was fairly different from others then in operation; he set up two separate masses and had them linked by piezoelectric transducers, used to convert any vibrations into an electrical signal. Each mass weighed about six hundred pounds. He and his colleagues conducted a seven-month run and saw no sign of a signal. "My aim," he told the conferees, "was not to try to find out if Weber's work was right or wrong but to find out more about gravitational waves." Meanwhile, Weber used his time to defend his methods, answering his critics over computer programming, his choice of algorithms in analyzing data, and the calibration of his detectors.

At the end of the session, Drever pondered the meaning of the differing results. "You have heard about Joseph Weber's experiments getting positive results. You have heard about three other experiments getting negative results and there are others too getting negative results, and what does all this mean? . . . [Is there] any way to fit all of these apparently discordant results together?" There were potential loopholes. Drever's experiment was not sensitive to long pulses. Perhaps that's why he was missing a signal. Kafka and Tyson were sensitive to certain waveforms, the ones most expected by theorists. But maybe nature wasn't acting according to the script. And yet, Drever's design would have caught unusual waveforms, since it could capture a broader range of frequencies. "I think that when you put all these different experiments together, because they are different, most loopholes are closed," Drever concluded.

For some the Tel Aviv conference marked the end of their involvement in gravitational-wave physics. Their interest swiftly waned when Weber's findings were put into doubt. But the field was hardly reaching an endpoint. To the contrary, the atmosphere in Tel Aviv was one of great excitement and eager expectations. Other participants felt that they had only scratched the surface of the field's potential. Many were excitedly talking about the possibility of increasing the sensitivity of the bars to detect supernovas out to the Virgo cluster of galaxies, some fifty million light-years away. So instead of waiting for a Milky Way supernova to go off every thirty years or so, they would enlarge their territory and possibly catch several events a year. To do that, however, sensitivities had to increase a million to ten billion times over 1974 standards. But such a sizable leap was hardly deterring the newcomers to this new brand of astronomy. Now that they had dipped their toes into the water, they were ready to plunge in headfirst.

Nearly everyone was coming to agree that Weber was mistaken. "We are not certain [that Weber is wrong], but it is probable," noted Drever in his concluding remarks in Tel Aviv. But he stressed that, even with null results, the field had broadened. People were beginning to bubble with ideas on what else could be done to improve their chances of seeing a bona fide wave. Some were choosing the low-temperature route. Others thought of using crystals with large resonances, enabling the detectors to ring for longer times. "Another technique which is coming into view now," said Drever, "is the quite different possibility of having separate masses which are a long distance apart. . . . One may monitor the separations using laser techniques."

But their desire was more than to just push the technological envelope. Science was always on their minds. "From a confirmed initial discovery that can be reproduced readily, I think the thing could rapidly spread to where we would have a real astronomy and we would be producing maps of the sky of gravitational wave sources," said Drever. "Every time we have looked into the sky with a new kind of detector, a new black box, we have found something which we did not expect," added Tyson.

In later years Weber occasionally had his hopes built up. A week's worth of data from the University of Rome, taken with a supercooled bar in July 1978, were compared to Maryland data taken at the same time with one of the room-temperature bars. In a 1982 paper in *Physical Review D,* a claim was made that correlations were seen between the two detectors. Weber immediately announced that this was additional confirmation that he and others were seeing gravity waves. The Rome researchers, however, preferred to describe it as a "background." Their conclusion in the paper was cautious: "The observation of a small background of coincidental excitations tells us nothing concerning the origin. Detection is statistical. There is no way of separating the coincidences which are due to chance and those due to external excitations. And we cannot be sure what fraction of the external excitations are of terrestrial or nongravitational origin."

Despite their doubts over Weber's methods, even his harshest critics recognized the tremendous engine that the Maryland physicist had set into motion. "It is clear," said Tyson in 1972, "that if it were not for Weber's work, we would not be as near as we are today to the possible detection of gravitational radiation ... with the use of supersensitive, low-noise antennas." Weber had created a momentum that could not be stopped.

Within ten years of the Tel Aviv conference, scientists had developed the second generation of bar detectors, each cooled with streams of liquid helium to a chilly −456° Fahrenheit, near absolute zero. This cuts down on the thermal noise inside a bar, which creates motions hundreds of times larger than a gravity-wave displacement itself. For a while one of the most ambitious of these projects was located at Stanford University, under the guidance of both William Fairbank and Peter Michelson. Their five-ton aluminum bar was situated in a hulking metal tank within a vast room that was once an end station for the original linear particle accelerator laboratory at Stanford. When conditions were good, the supercooled bar could at times detect a strain of around 10^{-18}. This meant that it could register a shiver that changed the bar's dimensions by one part in a billion billion, a ten-thousand-fold improvement over Weber's first instruments. The use of superconducting instrumentation, as well as the cooling of the bar, contributed to this sensitive performance. Such sensitivity put supernovas exploding all over the galaxy within their reach. Unfortunately, the 1989 Loma Prieta earthquake seriously damaged the Stanford detector. Too expensive to rebuild from scratch, the entire operation closed down.

But the work continued at Louisiana State University, which had been working hand in hand with Stanford and had a detector identical to the Stanford bar, ten feet long and three feet wide. In 1970 William Hamilton, a protégé of Fairbank's, had arrived at Louisiana State to supervise the construction of both bars at a nearby NASA facility just over the border in Mississippi. He later teamed with Warren Johnson to run the gravity research effort at LSU. Around 1980 they switched to a new aluminum alloy with a far better resonance. Slimmer than their old bar at a mass of five thousand pounds, this detector was known by the musical name of Allegro, which stood for "A Louisiana Low-temperature Experiment and Gravitational Radiation Observatory." It was an apt moniker for an instrument listening for a gravity wave's tones, which happen to fall in the audio range. The bar was tuned to 907 hertz and at its best could detect a quiver smaller than 10^{-18} meter.

Allegro came on the air in 1991 and operated until 2008, when it was at last decommissioned. Over those years, banks of electronic equipment sat right by the tank and monitored the bar's every movement. If a wave did come in, it would have caused the ends of the bar to go in and out, ever so slightly. Those small movements were then transferred to a secondary resonator attached to one end; this caused the tiny motion of the big mass to translate into a bigger motion of the small mass. Nicknamed "The Mushroom," because of its shape, this resonator amplified the signal. Unless it was down for an upgrade, Allegro ran twenty-four hours a day, seven days a week, a significant achievement. Its operation served as an incubator for training a whole new generation of gravitational-wave experts. They never detected any gravity waves, said Hamilton, "but we [did] have some enigmatic noises." Added Johnson, "We [had] several events per day, things outside normal stochastic noise, but we [could] usually track them to local occurrences." On one spring day some pile drivers working down the street were detected in the computer record that was kept on the bar's output.

Allegro was not alone. Similar ultracold bars were built and operated around the world for many years. Together, they formed a gravity-wave detection network. The frequencies to which they were tuned ranged from roughly 700 to 1,000 hertz. The LSU detector was joined by the Nautilus in Frascati, Italy, near Rome; the Auriga at the Legnaro National Laboratories near Venice; the Explorer at CERN (European Center for Nuclear Research) in Switzerland; and in western Australia the Niobe (so named because it was made of niobium, a metal more resonant than aluminum. Indeed, a niobium bar can ring for several days once activated). All were capable of seeing either a supernova go off or two neutron stars merging within our galaxy. While up and running, they attempted to track down and comprehend all possible sources of interference: electromagnetic disturbances, cosmic rays, and seismic tremors. The Nautilus, cooled to within a tenth of a degree of absolute zero, has actually recorded the vibrations generated by sporadic bursts of energetic particles passing through the bar, showers created whenever cosmic rays strike Earth's atmosphere. Once such noises were understood and subtracted out, bar researchers then focused on the unexplained events and conducted data comparisons. Allegro and Explorer, for example, had their data compared from a 107-day period in 1991, although no coincidences were revealed.

If major bar detectors remained in operation, they would have had a chance at seeing a nearby supernova, were the explosion strong enough. As luck would have it, though, no second-generation bar detector was on the air on February 23, 1987, when the light from Supernova 1987A, which had exploded in the Large Magellanic Cloud right here in our galactic neighborhood, at last arrived. And, on average, astronomers get to see such nearby events only every thirty to fifty years. All the advanced, supercooled bars at the time were temporarily offline to make improvements.

But certain room-temperature bars were working. Although Weber by then had ceased publishing his data regularly, he did occasionally surface to make a report. At the American Physical Society's spring meeting in 1987, he announced that one of his detectors registered excess noise—vibrations stronger than background—for a few hours around the time of the explosion. A gravitational-wave detector operated at CERN by University of Rome physicists also reported seeing events at that time. The alleged pulses occurred over a period during which particle detectors situated in Italy's Mont Blanc, in an Ohio salt mine, and in Japan recorded an extra burst of neutrinos. The probability that such a coincidence could happen, claimed Weber and the Italian researchers, was one in one thousand to ten thousand.

Was the detection real? The rest of the community is nearly unanimous in saying no. Other physicists have closely examined the data and have concluded that the Maryland-Rome statistics are highly flawed. For one, the Mont Blanc neutrino detection remains an enigma. The neutrino detectors in the United States and Japan saw the supernova go off four and a half hours later than Mont Blanc. It's hard to understand how a supernova—a violent and abrupt event—could have persisted for several hours, as the Maryland and Rome bars seemed to suggest. The Maryland-Rome collaborators even reported, "It is possible that new physics [is] needed," to explain the origin of the correlations.

That was the end of the road for Weber. He lost his longtime support from the National Science Foundation shortly after the announcement of his supernova claim. One gets the impression that if Weber had not taken such a hard line over those years—refusing to consider the possibility that his data might be spurious—he would have remained an honored member of the gravitational-wave detection community rather than moved to its sidelines. "It was one of Weber's failings, but a fatal one," said Saulson. "He would stick to his guns, despite all the mounting evidence."

By 1998, two years before his death at the age of eighty-one, Weber's hair had thinned and whitened, but he was no less a striking presence with black glasses that framed intense blue-gray eyes. For visitors he dressed formally: gray suit, white shirt, and solid red tie. He was always an avid jogger and mountain climber, boasting that he had climbed all the fourteen-thousand-foot-high mountains in Colorado. Even in his late seventies he was still trim, having maintained his Naval Academy weight, and he continued to be vigorous and quick to defend. Knowing that a reporter was about to visit, he amassed an organized defense, laying out his argument paper by paper in chronological order. Textbooks now take a uniform stand on his work; all state that his gravity waves were never confirmed. He picked up a representative book and recited aloud the offending section: "Other scientists have built more sensitive instruments and yet gravity waves have not been detected." He was clearly dismayed.

Weber long held the title of senior research scientist at both the University of Maryland and the University of California at Irvine. He liked to say that he was fired when he reached retirement age at seventy, but he continued to keep watch over his experiment when he could. His first wife died of a heart attack after twenty-nine years of marriage. In 1972 Weber married astronomer Virginia Trimble and began the bicoastal arrangement, spending part of the year in Irvine, where Trimble has a post, and the remainder in Maryland. His Maryland office looked more like a storage closet, filled to the brim with seventeen file cabinets that had been pushed together at the center of the room. Boxes of papers and bookcases lined the walls. Only one tiny aisle was left open to reach his small desk, situated against the far wall by the blackboard. Anyone dropping by had to sit knee to knee with Weber. Placed prominently over the desk was a picture of Einstein, one that Weber was told was Einstein's favorite portrait. It is a serious pose of the physicist as a young man that best accentuates his compelling eyes.

Weber was obviously disappointed at his inability to catch the brass ring. First with masers and later with gravity waves, he was the man who would be king. A scientist once prone to volatile outbursts at physics meetings, Weber displayed no anger during his interview, one of his last. His voice and manner remained matter-of-fact as he discussed his point of view. He conceded that his original estimates of signal strengths, based on classical physics, were wrong, since they implied energies being emitted in our galactic core that were hard to imagine. But that did not mean there were no gravity waves, he stressed. He had since devised a new way to think about how his bars were receiving a signal. "When our work first started, it was based solely on Einstein," said Weber. "But around 1984 I started asking how Niels Bohr would have done it." In other words, he wondered whether quantum mechanics—how his bar was receiving the signal on the atomic level—might better explain what he was seeing. Just as classical theory cannot explain how a metal can superconduct (that is, allow electrical current to travel with no resistance) or exhibit the photoelectric effect (a light wave acting as a particle), argued Weber, so did classical theory fail in explaining how a gravity wave interacts with the bar. He claimed that a gravity wave can couple to individual atoms and, as a result, interact far more strongly. This, he said, explained how he could see events a billion times fainter than once thought: "My theory says that a bar is made up of 10^{29} atoms, coupled by chemical forces and described by quantum mechanics. It's no longer a single big mass." The atoms, according to this scheme, all work in concert with one another to amplify the signal, making it a billion times stronger than the older theory could account for. According to Weber, the excitation bounces from end to end, with his center electronics best seeing the energy with each pass. Other bars missed this effect, he contended, because their sensors are situated on the end. Few physicists, though, agreed with this hypothesis. Weber wondered whether some of the groups contesting his claims spent enough time to confirm his findings. "If your objective is to show that the effect is null," he said, "you don't have to spend five years trying to get the experiment to produce a null result. You turn it on, and it produces a null result. They didn't give it enough time, enough care."

Tyson, who kept his hand in gravitational physics by going on to find and study gravitational lenses throughout the universe, applauded Weber's ingenuity. "Joe came up with a fantastic idea, which, to this day, is state of the art for bar detection," he said. "You have to praise Weber for figuring out how to do that." Where they parted company was in the interpretation of the bar's response. Tyson has concluded, along with others in the field, that Weber misunderstood the exact nature of the natural noises emanating within a bar, which led him to interpret what are really false alarms—simultaneous but random vibrations in the bars—as coincident signals.

Weber's observatory was still operating in 1998. He drove to it in a 1972 powder blue Volvo (his wife's). The lab was situated at the edge of the campus in a thick grove of trees, near the college golf course. The boxy white structure could easily have been mistaken for a small garage. Inside, a small overhead bulb cast a dim light. One detector was half hidden behind a jumble of odds and ends. It resembled a giant red oil tank. A second detector was about twenty feet away against the back wall, looking like an aluminum water heater. The aisle was too cluttered to reach it easily. The state of Maryland continued to pay for the electricity and special environmental controls to maintain a stable temperature in the room. When asked about other expenses, Weber silently answered by putting his hand into his suitcoat and pulling out his leather wallet. He did receive a NASA grant at one point, which allowed him to purchase a computer to keep track of his data. His bars were continuing to operate twenty-four hours a day, and a data point was registered every tenth of a second. Weber paid for the storage disks himself.

Weber leaned over to turn on the computer monitor screens, whose vibrant colors added the only touch of modern technology in a room filled with aging equipment. Two jagged lines—the ongoing signals from each detector—silently scrolled from right to left. The red tank contained a cylindrical bar two feet wide; the gray tank at the end of the room held a fatter bar, thirty-eight inches wide. Both had been running since 1969, three full decades. Weber admitted that his bars were deteriorating. The mounts were worn, and sensitivity was down. But he was determined to keep them running somehow until the latest gravitational-wave observatories were online. He was sure their data would coincide. "This observatory is important," he said, pointing to the two detectors, "because it's producing data." Judging those data, however, was always a matter of great disagreement—between Weber and nearly the entire gravity-wave community that he spawned.

Inspired by a suggestion from his astronomer wife, Weber went to NASA in the mid-1990s to suggest that the signals from his ever-working bars be compared to data received from BATSE (Burst and Transient Source Experiment), an all-sky monitor mounted on NASA's Compton Gamma-Ray Observatory launched in 1991. BATSE has been observing gamma-ray bursts over the entire celestial sky. Receiving a ten-thousand-dollar NASA grant, Weber used it to pay for a postdoc to carry out the statistical comparison. Of eighty gamma-ray bursts registered by BATSE between June 1991 and March 1992 the postdoc found that twenty of these events coincided within half a second to pulses from the larger of Weber's bars. Weber claimed that the probability of this being coincidental was one in six hundred thousand. Some gamma-ray bursts are thought to arise when a newly formed black hole ferociously devours its surrounding gas, in the process shooting out an intense pulse of energy from its poles. Other bursts may occur when binary neutron stars collide, a star explodes, or a lone pulsar is hit with cosmic debris. "After the data were written up and submitted for publication, I took the list down to NASA," said Weber. "NASA identified eleven out of the sample as due to one particular bursting pulsar near the galactic center. There's every reason to believe that source has operated sporadically over the years." When asked why other gravitational-wave detectors—instruments far more sensitive than his own—were not registering the same waves, Weber simply shrugged and suggested that no one was allowed to discover a signal until LIGO was up and running.

This work on gamma-ray bursts fueled Weber's hope that he would eventually be vindicated. He admitted that it was hard to explain some of the signals he had originally been seeing. "It could have been noise," he conceded, perhaps even an atmospheric phenomenon that affected both separated sites. But for him the gamma-ray bursts offered a definitive cosmic source. "I'm sure of two things," said Weber. "Death and taxes. But the evidence that I've seen gravity waves is overwhelming."

NASA did not continue its support of Weber's data comparison, but the work was reported in an Italian journal, *Il Nuovo Cimento,* noted for its liberal policy of publishing controversial results. No one listened, and no one followed up. Like the reaction to the fabled boy who cried wolf once too often, the rest of the community simply ignored Weber's last work. They didn't believe that room-temperature bars, antiques by present-day standards of technology, could have been recording anything more than local noises. Outsiders were sympathetic to his tale; scientists in the field were less understanding. No one, though, will take away his historic stature. His first bar is archived in the Smithsonian Institution in Washington, D.C. Another of his original bars, one used in his experiments, is prominently displayed in the lobby of LIGO's Hanford observatory. The others are unaccounted for, possibly sold for scrap metal.

Dissonant Chords

RAINER WEISS IS A DERVISH, WHIRLING ABOUT his office at the speed of sound—his sounds. He's refreshingly direct and rakishly profane. One spring weekend in the midst of LIGO's construction, Weiss was hunched over his desk, worried about a graph displayed on his computer screen. It exhibited the growing density of gases leaking into the newly installed vacuum tubes at LIGO in Louisiana. "This was actually expected," he said offhandedly. "The same thing happened at the Washington site." His high-ceilinged office was then situated on the edge of the MIT campus in a ramshackle, three-story building known solely by its numerical designation: Building 20.

If buildings can pass on good fortune, it certainly did that with Weiss, providing him with a favorable shot at finding a gravity wave. The wooden structure was built during World War II to carry out military research on a new technology known as radar. After the war, Building 20 witnessed the founding of the modern school of linguistics under Noam Chomsky, the assembly of one of physics' earliest particle accelerators, the refinement of the Bose speaker, and Harold Edgerton's astounding stop-action photography. "The plywood palace," they called it. Weiss himself punched through its flimsy ceilings and walls to help build the most accurate atomic clocks in the world. Building 20 was supposed to have lasted just through the war. With its stained siding and peeling paint, it remained standing for five decades as field mice and squirrels occasionally ran along the pipes in the

MIT's Building 20 was the famous "plywood palace," where in the 1970s Rainer Weiss first devised the plans for a laser interferometer that became the seed of LIGO. *Courtesy of the MIT Museum.*

corridors. Weiss's research group was the very last to leave before the building was demolished. It was too environmentally hazardous to preserve.

Weiss was not eager to pack up. "My whole career was here," he said at the time of the move, with a sweep of his arm as he walked down the silent hallways with their well-scuffed wooden floors. All the rooms were completely stripped of fixtures and furniture. Only wires and pipes remained dangling from the walls. Weiss's pedestrian office, with its worn-out couch and linoleum floor, stood as a lone oasis. Long tables jutted out into the room, each tabletop piled high with files stuffed with papers. Surprisingly, Weiss could recite the contents of each stack by heart. He's a talker. Stories, explanations, remembrances all spill out, seemingly simultaneously. He's feisty yet self-deprecating at the same time, a man who doesn't hesitate to tell you the story of how he flunked out of MIT as an undergraduate because he paid more attention to a girl than his studies. Short with thinning gray hair and an assortment of glasses to see at different distances, Weiss is a workaholic who was supposed to cut back after a heart attack (but hasn't).

First and foremost Weiss is an experimentalist. If he hadn't become a physicist, he'd be an electrician. He plainly prefers hanging out with scientists who get their hands dirty rather than those from the chalkboard brigade. It makes his blood boil to hear of experimentalists described as foot soldiers, with the generals—the theorists—looking down from the hill. "That's what those bastards want you to think," he said, with typical bluntness in 1998. "They think they own the field. They don't. Ideas come as much from experimentalists as from theorists. They're not generals. They're jerks like the rest of us. A good book has to be written on general relativity from the experimental approach. Wait until we discover black hole events. To understand what's really happening there will take more than what the theory can now provide. We're going to turn general relativity into a science, not just a description."

LIGO is a completely different approach to gravity-wave detection, extending and advancing the endeavor begun by Weber. Although LIGO is very much a collaborative project, involving hundreds of scientists, if forced to name the founding father of the effort, nearly all point to Rainer Weiss. (You might conclude that fate judges it that way: Highway 63, which goes right by the LIGO observatory in Louisiana, has long been known locally as Weiss Road.)

Weiss was born in Berlin in 1932, just as the fragile Weimar Republic was about to face its tragic end. His father, a physician, came from a wealthy Jewish family but rebelled by becoming a communist and marrying a Protestant, a German actress. Foreseeing the political troubles ahead for Germany, the family moved to Prague. By 1939 they were one of the last Jewish families allowed to emigrate to the United States, largely because of the father's much-valued medical degree. Plopped down into the middle of Manhattan, the young Weiss began dabbling in mechanics. "I took things apart all the time," he said. "Motors, watches, radios. My room was constantly a mess, and I was always getting into trouble for it." His father, who went on to become a psychoanalyst, was a bit annoyed. As a cosmopolitan German, he was more engaged with the arts and humanities, drama and literature. Weiss's one concession to his father's tastes was an

Rainer Weiss as a young professor at MIT around 1970, just as he was initiating his interest in laser interferometry as a means to detect gravitational waves. *With permission of Rainer Weiss.*

interest in classical music, which he came to love as an adult and expertly plays on the piano.

By high school Weiss began fixing radios for friends and acquaintances, an enterprise that eventually grew into a business. As soon as the war ended, surplus equipment started being sold on the streets of New York. "Equipment was flooding the cities," said Weiss. "You could go downtown and buy the damnedest stuff. If you knew even a little bit, you could get the most modern technology—transformers and radar sets—for a pittance." He would either skip school or go down on Saturday and pick up vacuum tubes, capacitors, any electronic component then imaginable. When a huge fire destroyed the Paramount Theater in Brooklyn when he was sixteen, Weiss salvaged ten loudspeakers, then the state of the art. He refurbished and sold them. In fact, he became so successful at making and selling complete audio systems, a novel endeavor at the time, that he almost didn't go to college. His business was highly profitable. But he also had an intellectual quest: he wanted to figure out how to solve the problem of noise, to take the hiss out of a recording as the phonograph needle rubbed against the shellac of the record. He specifically chose MIT so that he could study to become an audio engineer and learn enough to find a solution. What he didn't count on in 1950 was MIT's strictly ordered universe, far different from his rough-and-tumble neighborhood in New York where he fixed gang members' radios to keep from being harassed. All the buildings were known by numbers. All the courses were listed by numbers. "Everything was numbers," he said. "It was completely wacky. I asked myself, 'Am I going to survive in this place?'" He almost didn't.

Bored by his engineering courses, Weiss went into physics, but his record was dismal. Hopelessly in love with a Northwestern University student, a musician and folk dancer, he barely attended classes his junior year, spending most of his time in Illinois. When she dumped him, Weiss returned to MIT. He passed his exams, but the department flunked him out anyway for nonattendance. Despondent that he might get drafted into the Korean War, Weiss started walking around campus to clear his head. Passing by Building 20, he looked through one of its windows and saw two guys screaming at each other. "One guy was down on the floor looking at a big brass tube. The other guy was perched near the ceiling, and he was adjusting something up there," recalled Weiss. "They were trying to find a resonance in a beam of atoms, but it was hopeless." The two were working in the laboratory of Jerrold Zacharias, the scientist who constructed the first robust atomic clock for commercial use.

Listening to the argument, Weiss decided that the combative pair needed someone who knew electronics, his specialty. Hired on as a technician, he made himself indispensable. He found a home in Zacharias's laboratory (which both startled and chagrined his physics professors). That is where Weiss learned to design experiments, to build, and to manage. He got technical training in welding and soldering. With all the lab's high-tech equipment, he felt like a kid in a candy shop. "It was a style of physics that engendered you to do things," recalled Weiss with obvious fondness.

Eventually, Zacharias bent the rules to get Weiss back into school. As a graduate student, Weiss worked with Zacharias in making more and more accurate atomic timepieces. He specifically worked on a new concept, the atomic fountain, an experiment conducted right in Building 20. The idea was to send a beam of atoms upward. Like balls thrown into the air, the atoms would eventually stop and return to Earth. Once the atoms were slowed down at the turnaround, it would be easier to measure their vibrations, the very crux of an atomic clock. The setup first ranged over one floor. Weiss sent a hundred million atoms upward; none were recorded as ever coming down. Weiss soon punched through the ceiling to send the atoms two floors upward, then three. He kept going higher and higher in an attempt to see at least some of the atoms, the least energetic ones, finally stopping and falling back.

Weiss worked on this device for three years, only to discover that all the atoms were moving far more energetically than anyone had anticipated. The atoms were shooting right out of the building entirely. Today, decades later, a successful version of the atomic fountain has been built using supercooled atoms. It's not three stories tall but rather one inch high.

But there was one favorable outcome from Weiss's failed experiment: he caught the gravity bug from Zacharias, who long imagined his atomic clocks being used to test general relativity. Zacharias was hoping eventually to place one of his clocks on a tall mountain in Switzerland and another in a valley some seven thousand feet below in order to measure the gravitational redshift. It would have been an early version of Vessot's Gravity Probe A experiment, only this time carried out on the Earth's surface rather than in space. Weiss started learning general relativity in anticipation. The experiment was never done, but by then Weiss was hooked on gravity and in 1962 traveled to Princeton, the hub of experimental relativity, for a postdoc with Robert Dicke.

Weiss worked with Dicke on building gravimeters to measure the Earth's unique resonances, the modes in which it rang when excited by a seismological event. His detector went off the rails during the devastating Alaska earthquake of 1963. After two years, though, Weiss was eager to return to MIT. He loved its legacy of experimental freedom. "You could think of something and then do the experiment within a couple of days," he recalled. "You could do that because there was 'junk' all around and people who knew how to use that junk. That's why MIT was such a pleasure back then."

As a newly appointed MIT assistant professor in 1964, Weiss decided to measure whether the gravitational constant was changing over time, a project stimulated by Dicke's alternate theory of gravity. This spurred Weiss to work on lasers, a necessary piece of equipment for his measurement. At the same time he also began looking into the "tired photon" theory, a hypothesis (now discredited) that cosmic photons lower their frequency by losing energy as they travel through space. This introduced Weiss to the technique of interferometry, the method of choice for such a test.

In the midst of this ongoing research, the educational officer of his department asked him to teach the course in general relativity. "'After all,' he said, 'you ought to know it,'" Weiss remembered. "I couldn't admit that I didn't know it. I was just one exercise ahead of my students." He had learned relativity as an experimentalist, not as a theorist, and so taught the course from that perspective. To understand the concept of gravity waves, for instance, he came up with a homework problem. He asked his class to envision three masses suspended above the ground, their orientation forming the shape of an L. One mass would be in the corner of the L, the others at each end. The assignment called for his students to calculate how the distance between the masses would change as a gravity wave passed by. Weiss understood that as a gravity wave moves through space, it doesn't simply compress everything in its path and then, as it passes, expand it again. Rather, it has a multiple effect. It does two things concurrently in different directions. The wave compresses space in one direction—say north-south—while simultaneously expanding it in the perpendicular direction—east-west.

Though a gravity wave is a distortion of space-time, it conserves volume. The phenomenon is somewhat akin to the squeezing of a balloon: press in on a balloon's sides, and the rubber will immediately bulge out from its top and bottom, in a direction perpendicular to the squeeze. Gravity waves impose a similar effect on space-time. If a gravity wave were to come straight down on an L-shaped setup, passing through the Earth, the masses in one arm of the L would squeeze closer together while the masses in the other arm would move farther apart. This distortion can be visualized by looking at the weave in a piece of cloth as you pull it along one dimension. The squares of the weave distort in just this way. A millisecond later, as the gravity wave continues onward, this effect would reverse, with the compressed arm expanding and the expanded arm contracting.

About a year after designing this homework assignment, Weiss came back to the problem and realized that he had delineated a doable experiment, especially after dipping into the scientific literature. Like Joe Weber, Weiss was inspired when he came across the work of the British theorist Felix Pirani, who in a 1956 paper briefly mentioned how light signals could be used to detect gravity waves. But now Weiss recognized that laser beams, bouncing back and forth between the masses at each end of the L in his homework assignment, could monitor the expand/contract flutters as a gravity wave passes through. Here was a completely different way to detect gravity waves. What Weiss was imagining was a modified form of the interferometer that Albert Michelson used in his attempt to detect the ether. (Many frequently ask whether a laser interferometer can truly measure a gravity wave. If a gravity wave alternately stretches and compresses everything in its path, wouldn't it also stretch and compress the laser beam as well, making it impossible to measure a change at all? The answer lies in remembering that the speed of light never varies in a vacuum. The length changes in the arms are real and are revealed by the fact that the light takes longer to travel in one arm while simultaneously taking a shorter time to travel through the other arm. It is better to think of the light being used as a clock, not a ruler, in measuring the change. For a more complete explanation I recommend reading Peter Saulson's article "If Light Waves Are Stretched by Gravitational Waves, How Can We Use Light as a Ruler to Detect Gravitational Waves?" in the *American Journal of Physics*, volume 65, June 1997.)

A continuous stream of light from a laser would enter the corner of the system and be split into two beams, each directed down an arm of the L. Mirrors affixed to the center and end masses would then bounce the beams back and forth. (Later, the mirrors themselves became the test masses.) The beams are eventually recombined, at which time they optically interfere (hence the term *laser interferometry*). The beams initially could be set so that their wave patterns are out of step. In this way, when added together, the two beams would cancel each other out. The peak of the light wave in one beam is added to the trough of the light wave in the other beam, resulting in a null signal—darkness. But if a gravity wave caused the masses to move, the two laser beams would travel slightly different distances. In that case, when the length of one of the arms changes the tiniest bit, the beams will be more in step and produce some light when combined. The properties of the

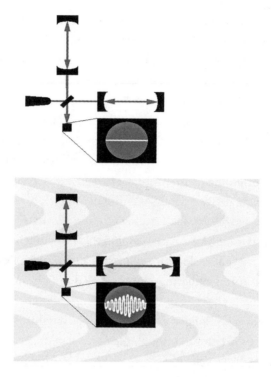

Laser light enters the system and is reflected in each arm to keep tabs on the length. When no gravity wave is present (*top*), the two arms of a laser interferometer are equal in length. When a gravity wave passes through (*bottom*), one arm contracts while the other expands, resulting in a signal. The movement here is highly exaggerated. The length changes are smaller than the width of an atomic particle.

gravitational wave are hidden within those light changes. Weiss came to see that the light should bounce back and forth many times before being recombined and compared, because the repeated ricochet would increase the total distance traveled and so magnify the difference enough for sensors to detect. The sensitivity would increase even more as the masses were moved farther and farther apart.

Weiss was actually rediscovering an idea that had been in the air but not widely discussed in the gravitational-wave community. Several years before Weber announced his purported signals, two researchers in the Soviet Union suggested using an interferometer instead of a bar to detect gravity waves. But the article, published in a Soviet journal in 1962 by Mikhail E. Gertsenshtein and V. I. Pustovoit, received no international notice whatsoever. Others in the field were completely unaware of its existence until Kip Thorne uncovered it many years later. "Gertsenshtein was the ultimate Caspar Milquetoast. He was unbelievably shy and mild. He had a number of seminal ideas that were totally ignored by the world," said Thorne. Weber independently thought of the idea as well. Although he did not publish it, he did discuss it with his protégé, Robert Forward, who drew a rough sketch of Weber's scheme in his lab notebook. And then Weiss came up with the concept on his own, motivated by his classroom exercise.

The interferometry idea became more and more popular over the years because, as the field evolved, it became quite apparent that bars had a number of limitations. The supercooling, for one, can be tricky. If something goes wrong, it can take several months to warm up the detector, fix it, and cool it back down again. And a bar's size limits the range of signals it can pick up. Fixed in length, a bar antenna can be tuned to only one frequency. If it were an optical telescope, it would be seeing only one color and no other, which limits its view. For many of these reasons, researchers began to focus their attention on the more versatile laser interferometer, which has the flexibility to carry out long-term astronomical investigations. It registers not just one frequency but a sizable range, increasing the chances of seeing a gravity-wave event.

A NASA astronaut served as the catalyst in getting Weiss's idea out of the classroom and onto the drawing board. In 1967 Philip Chapman had received his PhD at MIT in instrumentation, with a focus on general relativity. Going on to become a scientist-astronaut, he was on the lookout for gravity experiments to conduct in space and consulted Weiss, who had been on his thesis committee. "We were going to the Moon, NASA had plenty of money, and anything seemed possible," recalled Chapman of that era. Weiss told Chapman about his idea to use laser interferometry to search for gravity waves. Chapman himself had been thinking of approaches other than bars. Enthused by the prospect and looking for further collaborators in industry, he then talked to Forward, who was continuing his bar work at Hughes. (He had three small bars operating up and down the coast of California, one of them in his bedroom closet.) "Phil Chapman put me on to Weiss's interferometer idea," said Forward, resurrecting the notion he first heard from Weber years earlier. The possibility of a future NASA project got both Forward and Weiss working on the idea independently.

Weiss's role was highly significant in this venture, for he presciently envisioned more than four decades ago nearly all of the crucial pieces of the laser interferometer observatories that have now come online. Carrying out an extensive design study through 1971 and 1972, the first serious examination of the technique, he identified the fundamental sources of noise that gravity-wave observatories have to manage. Moreover, he completely outlined the approaches needed to control such noises. "I was trying to be like Dicke, who would initiate an experiment by first sitting down and thinking it through completely," said Weiss. Weiss's thorough analysis, published as an MIT Quarterly Progress Report, is now viewed as a landmark paper and is still consulted today. In an amusing coincidence, his career began with a determination to get rid of the noises in a hi-fi system, and he then transferred that interest to reducing the noises that could mask a gravity wave, whose wavelength happens to be in the audio range.

Meanwhile, Forward began to construct a small prototype. He and his colleagues, Gaylord Moss and Larry Miller, spent three years building and enhancing their system. The interferometer was located in a basement room at the Hughes Research Labs in Malibu, California. Two aluminum pipes, usually used for irrigation in farming, were set at a right angle and served as the laser beam tubes. Each arm was two meters (six-and-a-half feet) long and aligned to be most sensitive to radiation emanating from the galactic center (where Weber's signals were then supposedly originating). The masses, set in the corners of the L, each weighed a couple of pounds. The entire system was mounted on a granite slab manufactured by a gravestone company, which was set on air mounts for cushioning (an earlier setup rested on an inner tube). That was their greatest problem: isolating the instrument from various acoustic and ground noises. It was designed to receive gravity waves over a wide range of frequencies, from 1,000 to 20,000 hertz. As Forward pointed out in a journal article, they hoped that widening the bandwidth would give "significant insight into the nature of the source."

This tabletop system operated for 150 hours during the nights and weekends from October 4 to December 3, 1972, a time when the center of the galaxy was in prime alignment. The night shift had to be used because of the high levels of noise in the lab during regular business hours. Data collection was fairly tedious and required extraordinary effort. The researchers had to sit nearly motionless for hours at a time as they monitored the interferometer, so as not to introduce any extraneous noises. The output was recorded on a stereo tape recorder, and they listened in with earphones. "Gaylord Moss and I took turns spending the night 'observing,'" noted Forward. "I found it helped to keep my eyes closed and think as if I were part of the apparatus." One channel recorded the photodetector's output, while the other channel was used to monitor environmental interferences, such as noises in the laser beam, motions of the floor, any clatter in the lab, or audible sounds from the power lines. And in the background, like a rhythmic metronome, was the incessant *tic, tic, tic* of the National Bureau of Standards time signal, broadcast by radio station WWV. This was to make sure that any potential event could be timed to the nearest thousandth of a second.

At times various tones and clicks could be heard rising above a continual white noise hiss. Most of these sounds could be traced to either noises from the laser or thermal and mechanical contractions in the equipment. But occasionally, about once every ten minutes, there would be a distinct chirp from the interferometer that could not be traced to any internal noise or outside disturbance. None of these signals, though, were picked up by bar detectors operating at the same time. "In view of Weber claiming to have seen gravity-wave events," said Forward, "I believed it was worthwhile to operate the interferometer as an antenna for a few months, just to see if there was anything there. I did, and whatever leftover noises I heard on the interferometer were *not* Weber events." Chances were extremely slim that such a small prototype, the first of its kind, would have detected a cosmic signal in any case.

To improve its response, Forward had plans to take his interferometer to a remote site and extend its arms to much longer lengths, possibly a half a mile or more. An optical telescope gains more resolution and sensitivity by going to bigger and bigger mirrors to gather more photons. A laser interferometer gains sensitivity by extending the length of its arms. The expansions and contractions of space-time are simply easier to discern as longer and longer distances are used because the effect is cumulative. If the mirrors are twice as far apart, they will move twice as much relative to each other when a gravity wave passes by. So, the longer the detector arms, the more intense the jiggle, which makes any signal more measurable. Forward imagined that eventually two instruments should be built on opposite sides of the country. But by the end of the tests at Hughes, he had exhausted the funds his company was willing to spend on a gravity wave telescope. And in 1972 Chapman had left the astronaut corps, which meant that funding from NASA to expand Forward's prototype was also not available. Consequently, the Hughes Laser Interferometer Gravitational Radiation Antenna project came to an end. But others continued the advancement of this new approach. One of the most innovative of these researchers was Glasgow physicist Ron Drever.

Drever's interest in gravity was sparked around 1959, just a few years after he received his PhD in nuclear physics from the University of Glasgow. He came up with an intriguing way to test Mach's principle, the suggestion made by Ernst Mach that inertia, the tendency of an object to resist acceleration, arises when a mass interacts with all the other masses in the universe. With that hypothesis in mind, it was plausible to assume that a particle would accelerate differently toward a large collection of mass, such as the center of our galaxy, than at right angles to it, in a direction where mass is more sparse. That's what Drever tested. The particle in his case was a nucleus of lithium. When excited by a magnetic field (for this test the natural magnetic field of the Earth), the lithium could be made to produce an electrical signal at a specific frequency, a distinct spectral line. "I watched that line over a 24-hour period as the Earth rotated. As the axis of the field swung past the center of the galaxy and other directions, I looked for a change," said Drever. A change would indicate that the lithium was indeed moving differently, depending on whether it was directed toward or away from the massive galactic center.

Others had done similar experiments, but Drever (like so many physicists, a gadget lover since childhood) did it in a very offbeat way. He put together car batteries and assorted odds and ends in his backyard garden and ran the test from there. It was hardly makeshift, though: his experiment could have detected a shift nearly as small as one part in a trillion

trillion. "It beat everyone else who was trying to do it with much fancier stuff," noted Drever proudly. In the end he detected no change at all, at least to the level that he could measure. Inertia seems to be the same throughout the universe, no matter where a mass is headed. Such tests in physics are now known as Hughes-Drever experiments. Yale physicist Vernon Hughes independently conducted a similar test at the same time. Afterward, Drever spent a year at Harvard, where he constructed sensitive radiation detectors for Robert Pound's gravitational redshift experiments.

Through the 1960s Drever specialized in building detectors for nuclear physics and other applications. He also dabbled in cosmic-ray physics, studying the light emitted as the cosmic particles raced through the atmosphere. During a visit to southern England to conduct these tests, Drever stopped by Oxford University to hear Joe Weber lecture about his recent claim to have discovered gravity waves. Drever immediately thought, "If he's right, I'm sure we can do better than that," which brought him into the infant field. He and his group in Glasgow eventually got two bar detectors operating—but ultimately saw nothing. "Weber was wrong, and I was very sad," remarked Drever. "I was hoping he was right, because then we'd be in business."

Having no experience in cryogenics, Drever figured that he couldn't compete with the supercooled bars then under construction at Stanford and Louisiana State, so he chose a different path. Forward had recently visited Glasgow and had talked with Drever about his pioneering experiment in the Malibu basement. "I thought that the interferometers were likely to be better in the end and also much cheaper," said Drever, a prime consideration in Scotland, where funds for novel projects were scarce. One staff member was good at hunting down local companies that could make things cheaply. The vacuum tanks for their first bars had been made by a firm that manufactured ovens and other food industry equipment.

Using secondhand parts and equipment from his earlier bar experiment, Drever's group constructed its first interferometer in 1976. The one expensive item was the laser. But this new venture didn't go as planned. Drever quickly learned that laser interferometry was going to be far more difficult than he had initially imagined. The first problem that cropped up was simply light scattering. As light bounced back and forth between the mirrors in the interferometer, with each reflection tracing its own glowing thread as it hit a different part of a mirror, much of that light got lost. It scattered off imperfections in the mirrors. Drever's solution was to switch from a Michelson interferometer to a Fabry-Perot interferometer, a scheme that allowed the light over its many round-trips to stay as one beam and confine its reflections to a small area of each mirror. This reduced the chance that the light might ricochet off a bump on the mirror and go off in the wrong direction to ruin the measurement. It also increased the light efficiency tremendously. "The big advantage to me as a Scotsman was that this design was much cheaper," said Drever with a wry smile. That was because, using this scheme, the mirrors could be made smaller, the vacuum pipes as well. That was decisive for the field's advance. The technology did not yet exist to polish particularly big mirrors to the fine levels of smoothness required in this endeavor.

But there was also a downside to this new design: it would not work unless the laser was extremely stable, far steadier than any laser yet available. At that time the wavelength of light put out by the laser would sporadically jitter, which would have made it impossible to keep track of the infinitesimal shivers in the masses mounted in a gravity-wave detector. Undeterred, Drever simply invented a means of keeping the wavelength of a laser's light pure and steady. He later discovered that the idea was similar to one Robert Pound had used earlier for microwave cavities. Drever figured that a laser could be stabilized—its frequency kept fixed—through a feedback mechanism, locking the laser in a special way to an optical cavity. He visited John Hall, a leading laser expert at the Joint Institute for Laboratory Astrophysics in Colorado, to build such a stabilized laser, since the institute had ready access to the necessary parts. Drever and his colleague James Hough also built a cruder version in Glasgow. "The Glasgow one was comical," said Drever. "It was largely contained in tobacco tins. At that time Jim used to smoke a pipe a lot, and so he had dozens of tins around. They made good screening for the circuitry. The device used about a dozen tobacco tins."

Just as Drever in Glasgow, Weiss at MIT, and a seminal group in Germany were beginning these investigations into laser interferometry, Kip Thorne was working on the theoretical end of this enterprise. He was making the theory of general relativity user-friendly in the search for gravity waves. It was the time when Thorne was a rising star at Caltech, working with his students in making general relativity more testable by generating sets of parameters that experimenters could measure. In the process Caltech was replacing Princeton as a world center in relativity theory.

Thorne himself became involved with gravity waves in 1968 when he was introduced to Braginsky. Immediately impressed by the Soviet researcher, Thorne began a collaboration. Until the end of the Cold War, Thorne spent about a month in Moscow every other year, becoming the unofficial house theorist for Braginsky's gravity-wave group. From Braginsky, Thorne became convinced of the long-term experimental possibilities of gravity-wave detection, although he was skeptical of any short-term success. His wary attitude soon changed, though.

The turning point came at a meeting in the medieval town of Erice, Sicily, a favorite summer spot for physicists to convene at the Ettore Majorana Center for Scientific Culture. Held in an old monastery perched on a cliff overlooking the Tyrrhenian Sea, the 1975 conference had been organized by Weber to take stock of the field and discuss advanced techniques. As a theorist, Thorne had been calculating the gravity-wave strengths expected from various astronomical sources. Listening to the experimental presentations, he began to see that researchers had a good chance of getting to the required sensitivities with advanced techniques, such as the use of special materials or going to low temperatures. "I came away from that meeting convinced that the field was very likely to succeed," he said. "I hadn't had that kind of conviction before. It was because I was seeing the ideas for improvements in the detectors and what one might plausibly expect those ideas to achieve over ten- or twenty-year timescales." As a result, Thorne became the field's most dedicated barnstormer, going around the United States giving talks about the field's promise and the sources that might be detected with the new technologies coming online. "The Weber controversies had left a black mark on the field to some degree," noted Thorne, "and there was the need to erase that and maintain momentum in the United States."

Thorne was instrumental in convincing the Caltech faculty and administration to establish a gravity-wave detection team at the university, a natural complement to his theory group. Thorne was not wedded to any specific approach at first. "My attitude," he said, "was to leave it up to whomever we hired to decide the best direction." Weiss, though, hoped to change his mind. Weiss was chairing a NASA committee in 1975 on relativity experiments that the agency might carry out in space. He was already chairman of the science working group on another big NASA project, COBE, the Cosmic Background Explorer satellite then being built to measure the vestigial hiss of the Big Bang with exquisite precision. Along with his gravity-wave investigations, Weiss has been a major player in ongoing measurements of the microwave background, first with balloons and then from space. He was one of COBE's originators.

Because of Thorne's expertise in general relativity, Weiss invited Thorne to come to Washington and speak to his NASA committee. Failing to make a hotel reservation, Thorne ended up bunking with Weiss, but instead of sleeping the two stayed up until dawn in conversation about gravity-wave detection. Thorne at the time didn't hold out much hope for laser interferometers. In fact, in a section on gravity-wave detectors in his book *Gravitation,* Thorne had written, "As shown in exercise 37.7, such [laser interferometer] detectors have so low a sensitivity that they are of little experimental interest." That night Weiss, scribbling his proofs on hotel stationery, began to convince Thorne that laser interferometry was not a loser but instead a strong contender. (For many years, Weiss kept a copy of Thorne's quotation posted on his office door, just to tease the Caltech theorist whenever he visited MIT.)

After a committee study, Caltech agreed with Thorne to recruit the world's leading expert in gravity-wave detection, someone who would direct construction of a sophisticated prototype, either bar or interferometer, that allowed the university to refine the techniques and hardware necessary for a future gravity-wave observatory. Thorne would have liked to have brought in Braginsky, but with the Cold War still in progress that route was not possible. Braginsky feared the consequences to his family and colleagues. The transfer would have been viewed as a defection. And Weiss was then heavily involved in the early phases of COBE, which diverted his attention. But another name was often at the top of the lists of Thorne's consultants: Ron Drever. "Drever then had the best track record in terms of dealing with technical obstacles. He was preeminent. He had beautiful ideas, ideas that people would pooh-pooh at first and now they're incorporated into LIGO," said Thorne.

Drever's research team in Glasgow was just then beginning to improve its laser interferometer, employing Drever's latest modifications. The size of the instrument was fixed by the length of the available room, an old particle accelerator laboratory. The interferometer arms were ten meters (about thirty-three feet) long. "It was a struggle to make it all work," said Drever. It was in the midst of this effort that Caltech began to vigorously court Drever, who was ambivalent about moving as his Glasgow group was making progress. Although Caltech was "big leagues," as Drever put it, he was more attracted to the way in which European universities supported new endeavors. "I was quite happy where I was. You could do a lot with little money. The university employed technicians who could be used on any project. That meant you could try out new ideas without having it tied to some grant," noted Drever. By 1979, though, Drever finally decided to spend half of his time in California, which gave both him and the university the chance to see if the new Caltech venture was viable. It was. After five years Drever became a full-time faculty member. And with Drever on board, laser interferometry became Caltech's method of choice.

The growing momentum of laser interferometry had already caught the eye of the National Science Foundation. When Richard Isaacson arrived at the NSF in 1973 as associate program director for theoretical physics, he recalled his predecessor giving him a bit of parting advice: "I was visited a few weeks ago by a very clever guy—Rai Weiss," said Harry Zapolsky. "He has some interesting new ideas about gravity-wave detectors. If he comes back, you should pay attention." Eventually reviewing its national program of gravitational-wave detection at the end of the 1970s, the agency decided to expand its funding into this new arena. Also influential was Caltech's stepping forward and investing its own money in the technique. "Physicists tend to follow one another," noted Weiss, who had faced delays when he first approached the NSF for funding. "Once a large and prestigious university decided to go into it, that gave an extra little nudge."

With money from both Caltech and the NSF, Drever set up a full-fledged gravity-wave laboratory on the northeast corner of the university's campus. Stan Whitcomb, who became Drever's right-hand man, came on board to

assist with overseeing its construction. Caltech's aim was to build an interferometer identical to the one in Glasgow, only bigger. This interferometer
came to reside in a one-story structure, unassuming in its beige tone, that
wraps around a corner of a university laboratory, forming two long corridors. Only a modest sign on the door reveals the building's purpose. Inside,
the lab's most prominent features are two steel pipes (advertised as forty
meters long but actually thirty-nine, or about 128 feet) meeting at right angles. This length was not chosen for scientific considerations. Drever would
have gone even longer, but a tree was in the way, a tree that no one was eager
to cut down. A vacuum chamber stands at the corner of the L as well as at
each end of an arm. In each chamber a mirror/test mass is suspended. Each
mass is a cylinder of fused silica. (When first built, the masses were mounted in glass tanks christened Huey, Dewey, and Louie, after Donald Duck's
nephews—a tip of the hat to nearby Disneyland and typical of the school's
humor.) Inside the long pipes the laser beams reflect back and forth—you
might say from Dewey to Huey and from Dewey to Louie. Vacuum pumps
silently work in the background, keeping the pipes evacuated from stray
molecules that could disrupt the light's journey. To protect the suspended
mirrors from such outside disturbances as passing trucks or the seismic
tremors that occasionally shake Pasadena, the supports from which the mirrors are hung are cushioned by layers of stainless steel and rubber. When
first set up in the early 1980s, toy cars were used for the cushion, a colorful
assortment of tiny pink, green, yellow, red, and blue rubbery sedans. It was
a clever and cheap rubber source at the time but turned quite troublesome
in the end. Outgassing from the toys dirtied the vacuum system.

Since the early 1990s, the Caltech prototype has undergone several
upgrades, with a new vacuum system, new pipes, and new electronics
installed over the years. The goal was to make it a smaller version of the
full-scale observatories then planned for Louisiana and Washington. It continues to serves as a test bed for innovations. A graph on the wall depicts its
evolution. When first operating in the early 1980s, the prototype reached a
"modest" strain of 10^{-15} (able to discern a movement as small as a subatomic particle). By 1994 the system could measure a strain of 10^{-18}, a
thousandfold improvement. At one point it reached 10^{-20}. Above the coffee
maker in the lab, a row of celebratory champagne and whiskey bottles kept

track of these advancements. The progress came largely through a gradual but continuing series of technological improvements. Laser power, for example, has increased over the years, which directly affects the sensitivity of a laser interferometer system. The equipment is also better isolated from seismic disturbances. And, perhaps most crucially, the Caltech detector uses "supermirrors" as its test masses. Made of layers of dielectric material, these special mirrors lose while the interferometer is operating only several dozen photons for every million reflected.

As soon as he had arrived in California in the 1970s, Drever started checking out all the commercial vendors to find out who was making the best mirrors. He heard that Litton was making special mirrors for the military, for use in laser gyroscopes. They were not as yet available commercially, but Drever forged a connection and arranged for the company to make a special batch for his new interferometer. "They were fantastic, at least a hundred times better in terms of reflectivity losses," he said. With such mirrors in hand, Drever was spurred to think of additional improvements. When checking out his new supermirrors, he noticed that the reflected light was strong enough to bounce within the interferometer many times. Given such low light losses, he figured that he could recycle the light, have it bounce back and forth over and over again, which essentially boosts the power of the laser and increases the instrument's sensitivity. It was the development of the supermirrors, with their minuscule losses, which allowed Drever to even consider such power recycling. At the time this idea was revolutionary. "You'd catch the light and send it back in," said Drever. Now it's standard practice. When a visible-light laser is used, the effect can be spectacular. The laser beam enters the detector and reflects back and forth between the mirrors in each arm some three hundred times. It's as if three hundred laser beams are superimposed on one another. If the mirrors are aligned exactly right, so that the beams are in phase, the relatively dim laser beam suddenly brightens within the cavity into a brilliant shaft of light. With those two key improvements—a stabilized laser and power recycling—Drever enabled laser interferometry to turn a corner as a gravitational-wave detection system. He made the dream at last a possibility. With Drever's novel additions, the technique looked far more promising in its ability to reach the sensitivities needed to conduct astronomical investigations.

The Caltech prototype was never a true gravity-wave telescope. Today its team focuses no longer on increasing the prototype's sensitivity but rather on advancing interferometer technology—designing and testing advanced techniques and equipment intended for the LIGO interferometers. They're aiming to make control of the interferometer more automatic. But in the prototype's early days Caltech investigators couldn't help but try it out as a gravity-wave detector. For twelve days and nights in the winter of 1983, the Caltech interferometer was hastily put on the air after radio astronomers discovered a neutron star spinning what was then a record 642 revolutions per second, possibly jiggling space-time in the process. The 1987 Magellanic supernova, spotted in the southern sky, was examined too, though days after the initial burst. In both cases the Caltech detector perceived not a wiggle. According to Caltech physicist Rana Adhikari, who currently oversees the prototype, they won't be trying again.

From his previous experience as an experimentalist, Weiss had initially envisioned the field of gravity-wave astronomy growing steadily but very, very slowly. He figured that researchers in this arena would have to work on the innumerable technical challenges of laser interferometry before attempting a full-blown observatory. But a series of pressures and frustrations in the 1970s soon changed his assessment. Several years after Weiss returned to MIT's Research Laboratory for Electronics upon completing his postdoctoral fellowship at Princeton, its mission changed. Previously, federal grants to the lab could be applied to whatever ideas the lab was currently working on. Indeed, such funds helped him set up his first laser interferometer, a prototype with arms one and a half meters (five feet) long, and support the graduate students building it. But in the thick of the Vietnam War, a new rule was imposed which required that all research funded by the Defense Department, the lab's major funder, have a direct bearing on the military's needs. As a result, cosmological and gravity-related projects eventually lost their support. At the same time, Weiss was getting little respect from the MIT physics department, which was then more concerned with enhancing its solid-state division. "The faculty gave my students such a terribly hard time," he said. "They sneered at the extremely low sensitivity of the instrument." Indeed, in those early days Weiss found it difficult to convince many in the physics community that this new approach to gravity-wave detection had the potential to surpass the bars in sensitivity. Some thought the scheme was far too complicated—and perhaps even an erroneous method of detection.

As a last-ditch effort, Weiss arranged with the city of Cambridge to have the road right outside his laboratory—Vassar Street—closed down two weekends in a row at night. Stopping the trucks from rattling down the road gave his students the necessary peace and quiet to run their sensitive tests. Sawhorses were set up in the street to block the traffic. "We were trying to get my students a thesis," said Weiss. They obtained a strain of 10^{-14}, decent for a small prototype but impossibly weak for astronomical investigations. "During their oral exams," noted Weiss, "my colleagues had the temerity to ask these kids what they discovered. 'Well,' said one student, 'We didn't see the Sun blow up.' One professor replied, 'I can look out the window to see that. What do we need your data for?' They didn't see the technological aspects of it at all. That's when I decided I was never ever going to put a student in that situation again." With his funding slashed and his colleagues indifferent, Weiss realized he had to get a full-scale observatory under way. He needed to get into the business of astrophysics as soon as possible. And that meant going beyond tabletop detectors or prototypes with extended arms. It meant going big—*very big*. In 1976 he began to work on an idea that became the seed of LIGO.

Right off Weiss envisioned a system with two widely separated detectors. He initially set the length of the arms at ten kilometers, a little over six miles. Such a long length was necessary to make the instrument sensitive enough to detect the waves that theorists hypothesized were out there. Knowing that such a sizable facility would cost tens of millions of dollars, Weiss reasoned that he'd have to call an international meeting and try to turn its construction into a worldwide effort. He never imagined that he could obtain funding from the National Science Foundation alone, by then the only US funder for gravity-related research. "I knew right off that this was going to cost over $50 million," said Weiss. "It was outrageous to contemplate that the NSF would go into this field that (a) had that terrible history with Weber, (b) had no experience with big projects, and (c) had no scientific basis as yet. It was cuckoo."

But as soon as the NSF made its decision to fund the Caltech prototype, a sizable investment for the agency, Weiss decided to be more aggressive and offered his idea to leapfrog to an even bigger detector. Among gravity-wave researchers at the time, Weiss was the only person with a firm grounding in major physics projects, being in the thick of building the COBE satellite. From that experience he came to learn that it would be wise to build a large facility fairly quickly, and then develop and advance the techniques along the way. "Rai recognized, far more clearly than anyone else, that the only way to get to the required sensitivity was to build something with long arms," recalled Thorne.

The NSF gave Weiss the go-ahead to conduct a feasibility study of his ambitious scheme, then called the "Long Baseline Gravitational Wave Antenna System." Most needed were realistic cost estimates for constructing such a facility. Completed in 1983, in collaboration with Peter Saulson and Paul Linsay, the study (now familiarly known as the "Blue Book" for the

Richard Isaacson, the NSF's former program director of gravitational physics, in 1995. *Photograph by Tatiana Divens, courtesy AIP Emilio Segre Visual Archives, Divens Collection.*

color of its cover) ultimately convinced the NSF to initiate research and development. The decision came with one stipulation: any major laser interferometer observatory had to be a joint project between Caltech and MIT. There were both political reasons—the combined clout of two prestigious institutions to get the financial support from Congress—and technical reasons. Big interferometers required a much larger effort than you could mount with one professor and a handful of assistants. Weiss had anticipated such a collaboration forming all along.

With this decision, gravity-wave astronomy moved toward the big leagues. Weiss, Drever, and Thorne were in charge of overseeing the new collaboration. Given Thorne's Russian connection, the three came to be known as the "troika." The troika's first order of business was to assemble a detailed construction plan for two full-scale interferometers, which had been reduced to having four-kilometer (two-and-a-half-mile) arms owing to budget and engineering considerations. (The available sites, for one, were limited in size.) The NSF's response to their initial ideas, though, was decidedly cool. Their suggested schemes were judged not good enough to be viable, especially in an era when federal budget woes were putting the brakes on other big science projects. "Caltech and MIT were simply not ready. Their plans were premature," said Isaacson, who rose to become NSF's program director of gravitational physics and is now retired.

As a result, the troika went back to the drawing board to rework their plans. "Each step forward in the scientific community or the governmental funding process required the interruption of exciting progress in laboratory research or conceptual insights," recalled Isaacson. "Endless dog-and-pony shows, each demanded another week of preparation." During this period of turmoil, Isaacson and Weiss would occasionally meet, walking around Walden Pond outside Boston to deliberate uninterrupted. In the end, the troika continued to get lukewarm critiques. Normally, that would have been the death knell for a proposed science endeavor, but Isaacson and Marcel Bardon, then director of NSF's division of physics, had faith in the idea and pulled every string to keep the project alive. They made sure that money was provided to continue research and development on the proposal. "Bardon in particular recognized the technological promise," recalled Isaacson. "The intellectual excitement was overwhelming. But what

LIGO collaborator Peter Saulson of Syracuse University. *Courtesy of Peter Finger.*

we had to do was reduce the risks. We wanted to nurture the dream until Caltech and MIT were up to the job of managing such a project."

"It was a miracle," said Weiss years afterward. "So many things would have killed it, but the National Science Foundation was responsible for keeping it going." But as a result of this favored handling, the proposal became highly visible within the scientific community—attention that brought much criticism along with it. Richard Garwin, Weber's nemesis, began to loudly question the worth of building a large gravity-wave observatory so soon. He did not trust the grand claims being made for it by its supporters. With assistance from Caltech and MIT, the NSF answered by assembling a blue-ribbon panel, including Garwin, to advise the agency on "going big." The panel met for a week in Cambridge, Massachusetts, in the fall of 1986, bringing in the major players in gravity-wave detection from around the world to discuss the prospect. Also at the meeting were members from industry to discuss the technical feasibility of making the required optics, lasers, and servo systems. "It was the turning point for the field," said Thorne. "This hard-nosed committee produced in the end, after much internal debate, a unanimous report. It said that the field had great promise and the

correct way to do it was to build two big interferometers right from the beginning, because you can't get any science with one by itself."

But this support came with a strong caveat: urged on by Weiss and others, the committee recommended that the troika be disbanded and replaced
by one project director. There had been disruptive tensions between Weiss
and Drever all along, technical differences of opinion that made it difficult
for the two institutions to work together effectively. "It was five years of
sheer agony for everyone," said Weiss. Drever has always been driven by an
intuitive physical instinct. Weiss is far more analytical. Drever is essentially
a loner, while Weiss is experienced on big projects and more realistic about
the compromises required in such a setting. Drever preferred fine-tuning
the prototype before jumping up in size. Weiss was eager to build big, then
tweak from there. Thorne was caught in the middle. This mismatch in
temperaments became a serious problem for the project, holding up some
critical decisions. Consequently, the NSF demanded that a single director

From left, Kip Thorne, Ron Drever, and Robbie Vogt in 1990 at Caltech's forty-meter
interferometer, a test-bed for LIGO technology. *Courtesy of the Archives, California
Institute of Technology.*

be put in place with authority over the entire project. Caltech and MIT found that director in Rochus Vogt, known to all as Robbie.

Vogt took over in June 1987. He had a distinguished track record. Trained in cosmic-ray physics at the University of Chicago in the 1950s, Vogt served as the first chief scientist at the Jet Propulsion Laboratory in the mid-1970s, making science preeminent at the NASA facility. "He was also one of the best chairmen I have ever seen heading up the division of physics, math, and astronomy at Caltech," said Thorne. It was then that Caltech was building its millimeter-wave radio astronomy array in California's Owens Valley, which was in dire trouble during its construction and about to be canceled. Vogt went in and pulled it together on time and on budget.

But Vogt is also known as a tough man, who garnered political enemies over his career. As a youth in Nazi Germany, he developed a fervid distaste for wasteful authoritarian bureaucracies. With his short-clipped hair and black-framed glasses, he struck one at the time as a taller and leaner Henry Kissinger. Vogt was available only because he had just been fired as Caltech's provost, due to conflicts with the university president. Reluctant to accept the directorship at first (he yearned to get back to being a "real scientist who plots data"), Vogt eventually relented. University trustees sold the project to him as Caltech's next Palomar telescope, for many years the most powerful optical telescope in the world.

Vogt's organizational skills turned out to be invaluable to the project, which at last had an official name—LIGO. Vogt personally husbanded the final proposal. He got his fingers into all the nitty-gritty details and corralled everyone to get involved. He thought of himself as the "resident psychologist." Right off, he broke the scientific deadlock between Drever and Weiss by choosing Drever's Fabry-Perot design over the Michelson interferometer favored by Weiss. He even persuaded Thorne to turn experimentalist and come up with a successful solution to extinguishing stray light in the vacuum pipes. Vogt, who became LIGO's most ardent champion, was certain that if the project were terminated, it would kill gravity-wave astronomy for a generation.

The final proposal for LIGO got strong reviews at the NSF in 1990. An external review panel also gave it a thumbs-up. But the requested money was so sizable—an initial $47 million outlay of a total $211 million construction cost—that it had to be requested by the president and get Congress's okay. This was a first for the NSF. Unlike the Department of Energy, which regularly deals with large projects such as particle accelerators, the NSF had never before sponsored a project so large that it required line-item approval in the federal budget. Opposition immediately arose in the astronomical community, which proclaimed that such money would be better spent on telescopes. At the time $211 million was twice the total of the NSF astronomy budget. Astronomers were angered that the NSF was choosing to put the money into a gamble, rather than into already proven technologies. Was it worth such a high price to find a gravity wave, the critics asked Congress? Nobel laureate Philip Anderson, a condensed-matter physicist, wondered aloud, "If it didn't have Einstein's name on it, would you give a damn?"

LIGO researchers were particularly dismayed when a former member of the gravitational-wave detection community, Tony Tyson, testified against the project before a panel of the House of Representatives Committee on Science, Space, and Technology. He stressed to Congress that LIGO "demands a truly phenomenal increase of sensitivity," up to one hundred thousand times more sensitivity than the Caltech prototype was then attaining, before it could hope to obtain important astronomical data. A LIGO endeavor, to Tyson, was simply "premature," since too many engineering problems were still unsolved. He preferred a slow-but-sure approach. "Innovation is not necessarily synonymous with a 'shot in the dark,'" he concluded.

Berkeley astrophysicist Joseph Silk got a jab in while reviewing a popular book Thorne had written on general relativity. Silk wrote that LIGO "has misleadingly cloaked itself as an 'observatory.' Despite its name, any astronomy on the symphony of waves from merging black holes in remote galactic nuclei will have to await a greatly refined second-generation detector, and undoubtedly a vastly more expensive undertaking." Even bar detector scientists liked to point out that bars were cheaper and more developed. Others questioned the importance of gravity-wave studies altogether, insisting that scarce government funds were better spent on projects with lower price tags and far likelier scientific payoffs.

LIGO supporters countered that they were after more than sheer detection. They resented their project being called a shot in the dark. What they wanted, they said, was to open up a new arena for gleaning information from the universe, a method far different than gathering electromagnetic radiation. Including the word *observatory* in LIGO's title was a deliberate choice, an expression of their intention to use LIGO as an ongoing experiment. Moreover, they noted, LIGO's construction funds were independent of the NSF's regular budget. Rejection of LIGO did not mean that the funds would necessarily be applied to other science projects.

Even with blue-ribbon panels strongly supporting the project, Congress became wary when eminent scientists stepped forward to object. As a consequence, LIGO got stalled for two years. Vogt was a relative novice in his first dealings with Congress. In 1991 he failed to get the go-ahead for construction, although he did get funding for further engineering and design work. Concerns over the federal budget deficit were then high. Members of Congress questioned whether they were ready to invest such a large sum of money in an unproven facility.

By the next year, though, Vogt had honed his lobbying skills with the help of a consultant. He learned to be a better salesman when pitching the story of gravitation to key legislators. While in Washington, for example, Vogt wrangled a twenty-minute meeting with Louisiana senator J. Bennett Johnston, who later became an ardent, behind-the-scenes supporter of LIGO, especially when his home state was chosen as one of the two sites. "After I had my twenty minutes," recalled Vogt, "Johnston's senior staffer looked at his watch and said, 'Senator, the twenty minutes are up. Let's go.' But the senator responded, 'Cancel the next appointment.'" Vogt's tales of cosmology so captivated Johnston that he canceled the following appointment too, as well as the one after that. The two ended up sitting on the floor by the coffee table while Vogt drew pictures of curved space-time. Einstein's name once again wielded its magic power. In the end Congress appropriated the money.

With the funding came a sudden shift in the tenor of the project. The transition was quite evident at both MIT and Caltech. Blueprints and photographs lined the hallways. Memos cluttered the desktops. The offices more resembled an industrial corporation than an ivory tower. The initial laser interferometer team, when it started up, had been quite small. There were about a dozen people on each coast, including technicians, scientists, engineers, and administrative personnel. By the mid-1990s, there were more than 150 on staff, two-thirds at the Caltech headquarters and the rest at both MIT and the two detector sites. The jump from prototype to mature facility was a canyon-sized leap: the arms going from forty to four thousand meters, a factor of one hundred. "It was a big transition," said MIT's David Shoemaker, a deputy detector group leader for LIGO at the time. Scientists had to move from their individual lab environments, where they had total control, to a hierarchical facility. Participants now documented their every move and dealt with myriad outside companies. Once a singular pursuit, gravity-wave detection had become a networking of many players, each serving a specific role.

There were repercussions to this change. As with any burgeoning enterprise in science, LIGO has had its share of heated discussions, compromises (both political and scientific), and struggles between pioneering scientists with strong and volatile personalities. New fields often attract the risk takers, whose passions and fervor can be difficult to handle day after day. Some observers even label it hubris, the unwillingness to concede that someone else might see a better way to handle a particular problem. "Most of the conflicts could be blamed on the shift from tabletop physics to big physics," said Peter Saulson, who has been a LIGO collaborator since its beginning. "It's a time when you have to step out of your laboratory, set up deadlines, and follow budgets. Many of the original people didn't have this experience. Knowledge had to be brought in from the outside. At first we thought we were uniquely cursed. But I have since found out that it's part of every major science endeavor, where you invent something from scratch and transfer it to a larger arena." Construction of the two-hundred-inch Palomar telescope, for decades the largest optical telescope in the world, faced an identical crisis. Ronald Florence, writing on the telescope's long development, said that the scientists' "insistence on exploring, designing, and engineering every step of the project from scratch—doing basic research on subjects as varied as oil bearings, the wind resistance of dome sections, and the chemistry of glass—was for [the chief administrator] a sure route to a quagmire of indecision that would never see the telescope built. . . . Men who were building a unique machine, an instrument perfect enough to explore the secrets of the universe, didn't like that attitude."

A LIGO member with experience in industry saw this history play out once again. Research scientists, long used to laboratory independence—the freedom to change an experiment at will—became upset over LIGO's rigorous schedules and their inability to make last-minute changes in the instrument. Scientific considerations suddenly had to bow to financial and engineering concerns. This meant that certain technologies had to be locked in early, even though advances may have developed later. Some adapted; others left. One unwilling victim was Ron Drever.

Originally brought in to jump-start Caltech's entrance into gravity-wave detection, Drever was unceremoniously ousted in 1992 after growing conflicts with the LIGO team, particularly Vogt. "Leggo my LIGO" was the headline in the *Los Angeles Times* as it reported on the feud. A short, portly man with a bulbous nose and warm blue eyes—a Santa Claus without the beard—Drever was always talking and chuckling, ideas constantly bubbling up fast and furiously. And for some this rambling creativity posed a problem. Drever was most comfortable working in his own laboratory, where he could constantly alter his experiments as he developed new schemes. He was reluctant to lock into a final design if a better method was on the horizon. Drever was a veritable fountain of ideas, as many bad as good, but he would champion them all with equal energy.

And then there was Vogt, whose management skills played a large part in getting LIGO's approval. He was a man revered by many at Caltech for his organizational talents yet who was also feared for his sporadic outbursts and at times harsh tactics. He was in battle mode as he fought both a skeptical scientific community and a wary Congress for acceptance of LIGO in the early 1990s. You were either friend or foe to the LIGO venture, an uncompromising stance that some found hard to deal with. Ultimately, there was a clash of wills: Drever, always coming up with new approaches, wanting to try them out, versus Vogt, the manager wanting to maintain order and discipline because of a rigid time schedule. In a way the strain between them was also a conflict over ownership. Who got credit for LIGO? Drever, the brilliant experimentalist, whose major breakthroughs were recognized worldwide and enabled a facility such as LIGO to go forward, or Vogt, the expert strategist, who got the controversial project approved by Congress?

A review team, brought in to evaluate the LIGO project amid this internal strife in 1992 and 1993, likened the major players to "a dysfunctional family that needs to be split up." (One observer ruefully joked that Caltech needed "to put Prozac in its water coolers.") A university committee eventually agreed that Drever had been inappropriately dismissed, and Caltech provided money for him to continue his research on advanced interferometers independently.

With his settlement Drever set up his own forty-meter interferometer on the Caltech campus within a vast, cavernous room that once housed a synchrotron particle accelerator. One arm ran the length of the room, while the other arm went across the width of the building and out the end, where it continued in a tunnel beneath the road. Here for a number of years Drever tried out all sorts of new gadgets, like using magnetic levitation to suspend the test masses. "The people who do many wrong experiments are the best," he said. "You try out more things. That's how you make discoveries, but you have to be quick." There was certainly a quickness to Drever's step at this time, as well as in his speech and in his hands. He was always buzzing around, always thinking, always animated. He retired from Caltech in 2002, becoming a professor emeritus. He later moved to Scotland, where he was diagnosed with dementia and currently resides in a care home near Edinburgh.

Yet even after Drever left the LIGO project, tensions persisted. Vogt ran a tight ship and kept his decisions close to the vest, a managerial approach used at times on covert military projects. "Give me the money and stay out of my way" was his philosophy, a method he had successfully applied in building Caltech's Owens Valley radio telescope array. Vogt preferred working with a small elite team of scientists and engineers, free of governmental control and costly bureaucratic reviews. He was convinced that only then could LIGO be built for $220 million (the planned construction cost was ever rising). But the NSF ultimately decided that LIGO needed to be more open and accountable—its construction activities laid out in meticulous detail and a coherent plan developed for accommodating outside scientists. To do this the agency wanted LIGO's management to shift to a managerial mode long used on high-energy physics projects. Vogt resisted.

In response, after lengthy consultation with both MIT and NSF, Caltech replaced Vogt with physicist Barry Barish as LIGO's principal investigator in February 1994, just a month before groundbreaking took place in

Washington State for the first of the two LIGO detectors. Vogt, astute in the politics of science, had shepherded the project for seven years from proposal to funded experiment but was less prepared to deal with managing an evolving large-scale facility that was at last being cast in steel and concrete. Barish, though having no experience in gravity-wave research, was tapped for his expertise in handling large physics projects. The cancellation of the Superconducting Super Collider in 1993 made Barish available; he had been heading up one of its detector teams. Barish entered particle physics in its golden era during the 1960s, when new particles were quickly being found in great abundance. It was a time when universities were abandoning their in-house particle accelerators and starting to use large national facilities. So operating a facility like LIGO was familiar territory for Barish.

A longtime member of the Caltech faculty, Barish had watched LIGO undergoing its growing pains from the sidelines. He credited the initial team for performing the necessary conceptual work and pinpointing the technique's limitations. But, he added, taking the leap from tabletop physics to a miles-long facility was beyond their level of expertise. They had a certain blindness to this limitation, said Barish, even an arrogance when confronted with reviewers' criticism. They were literally founding a field, yet they hadn't grown up in the environment that particle physicists take for granted. The field of particle physics had evolved over decades. With LIGO, gravity-wave astronomy was growing up virtually overnight. Its evolution from small to big was occurring at a rapid tempo and this meant that certain scientific styles had to give way quickly to new approaches, ones more compatible with a large infrastructure. "In a small lab if you make a mistake, you can go in the next day and fix it," he explained. "But here, when you are committed to spending a hundred thousand or a million dollars, you can't fix it later. You need to have a system of checks and balances internally. In particle physics that's just part of the structure." When Barish took over the leadership of LIGO, just months after the Superconducting Super Collider was closed down, he found a battle-worn group recovering from the strains of the Vogt-Drever tempest. He set one simple goal: to build LIGO.

A Little Light Music

THE HANFORD SITE, A FEDERAL NUCLEAR PRODUCTION complex now under decommission and currently the nation's prime repository for nuclear waste, sprawls over hundreds of square miles of scrub desert within the rain shadow of the Pacific Cascade Mountains in south-central Washington. John Wheeler first came to this isolated spot in the 1940s when the government set up a manufacturing plant in Hanford to produce plutonium, an element that does not occur naturally, for use in the bomb that was dropped on Nagasaki, Japan. Settling in nearby Richland at the time, Wheeler remembered "a community of houses, stores, and schools that had been erected in a matter of months by the Army Corps of Engineers. The sidewalks were tar that had been squeezed out like toothpaste. Asparagus sprouted through cracks in the sidewalks from the farm that had been there before the Corps' bulldozers moved in." The town continues to thrive by the banks of the Columbia River.

It is ten miles from Richland to the Hanford facility, proceeding west along Route 240. The turnoff is not obvious. At the key intersection the road signs direct travelers to either continue west or turn south. There is no sign at all to explain the highway going north, the entrance to the reservation. It is a habit left over from Hanford's many years as a national secret. Five miles down that desolate two-lane road is a duplicate of the LIGO complex in Louisiana—the same cream, blue, and gray colors.

Standing alone on the vast plain, a landscape long ago carved flat by the immense outflow of an ancient glacial lake, the observatory resembles either a tasteful warehouse or a modern art museum inexplicably placed in the middle of nowhere. It is a rent-free guest on the Hanford Site. The observatory's nearest neighbors, though miles away, are a nuclear power plant and a mothballed research reactor. They share a vast, cloud-studded sky that stretches from horizon to horizon. Only to the southwest is this blue vault interrupted by the Horse Heaven Hills and the smoothly sculpted Rattlesnake Mountain.

Tumbleweeds, the dense Russian thistles accidentally introduced to the West, are ubiquitous. They continually roll over the barren terrain and pile up along the arms of the interferometer. "We baled two hundred tons of it last year," said Hanford observatory director Fred Raab in 1999, when the detector was under construction. "We have a crew out every week." The strawlike balls are gathered and baled like hay and then used for erosion control around the site. Otherwise, the roads along the miles-long arms would be completely blocked.

An aerial view of LIGO's Hanford Observatory in Washington state. *Courtesy of Caltech/MIT/LIGO Laboratory.*

Raab enjoys his job immensely. He must, because he's now directed the Hanford detector for more than twenty years. "Being first, you get to write the playbook," he noted. Raab joined LIGO when he was convinced that there would be gravity-wave detections in his lifetime. Considering his background, it was the ultimate challenge. Trained in atomic physics, he long dealt with precision measurements on the tiniest of scales. "To me a table is a bowl of Jell-O," he said, a reference to the constant jitters that occur on the atomic level. The gravitational jitters he captures with LIGO, though, are even smaller.

The journey begins in the complex's center station, an enclosed city of metal gleaming under bright lighting. One and a half million pounds of metal were used to construct the instrument. Seven hundred truckloads of concrete were brought in to build the floors and beam-tube roadways. The vacuum chambers, where the test masses reside, look like the tanks in a microbrewery, though unpolished. "It's more like a sewage treatment plant," said Raab with a laugh. But here, there is no noise. Only a whisper of wind from the air ducts. Gravity-wave detection starts with the laser, set in an alcove of the main hall. The laser sends its light into the vacuum system, where it is divided into two beams, each directed into a separate four-kilometer arm. As in Louisiana, the arms shoot outward to form the familiar L shape. In Hanford one arm is directed toward the northwest, the other to the southwest.

Given Hanford's legacy of covert operations, some locals initially harbored the suspicion that the observatory was a cover-up for a secret laser weapon project. They imagined that the arms were somehow hinged, able to rise up and fire a powerful ray blast into space. To demonstrate what is really happening in those miles-long tubes, Raab, an enthusiast for outreach education, built a small tabletop interferometer to take to local schools. Its light source, a toy laser pointer often used by lecturers, is held up by a clothespin. Its vivid red light is directed into a beam splitter, which creates two separate beams that are sent off at right angles. Each beam reflects off a mirror. On their return, the two beams are recombined and aimed at a white screen. When the two beams are "in phase," wave peak matching wave peak, a bright red spot is seen on the screen. When the two beams are "out of phase," the trough of one wave canceling the peak of the other (like +1 and −1 summing up to 0), a dark spot is seen. By pulling on a string to move a mirror ever so slightly, which changes the distance in one of the arms, the spot changes from bright to dark or dark to bright. The kids take delight in this effect. It is exactly what is happening inside LIGO. A tiny change of distance in an arm translates into a change of light intensity at the interferometer's output.

LIGO is a direct descendent of the Michelson-Morley experiment, whose failure in the 1880s to detect any variation in the speed of light led to the special theory of relativity. But where a detectable signal in that experiment would have been relativity's downfall, a verified signal in the LIGO detectors confirms one of general relativity's key predictions. Michelson was the greatest experimenter of his day. His interferometer was capable of detecting a length change equal to one-twentieth the width of a visible light wave. With LIGO today, researchers are able to do nearly a hundred billion times better. In terms of measuring a change in distance, it is currently science's most sensitive instrument.

The very nature of this endeavor appears paradoxical at first. To detect the very small—shifts in space-time smaller than a subatomic particle—requires

Two visitors gaze down one of the two four-kilometer arms at the Hanford observatory in eastern Washington State. *Courtesy of Caltech/MIT/LIGO Laboratory.*

an instrument that is very big, four kilometers long. But that's just the nature of measuring strains in space-time. To boost the sensitivity even more, researchers have applied an added trick: the laser beams go up and down the arms about 280 times, a total of some 1,120 kilometers (696 miles). This makes the arms act even longer, which greatly increases the chances of detecting a feeble gravity wave.

Working together, the Hanford and Livingston detectors form a single instrument that slashes diagonally across the continental United States. Separated by 3,002 kilometers (1,865 miles, or a hundredth of a second at the speed of a gravity wave), the two locations were among nineteen sites proposed in seventeen states. Officially announced in 1992, the final choice was made through a combination of politics, geographic considerations for signal triangulation, and scientific practicalities. The chosen areas, of course, had to be fairly flat on the surface, because underground tunnels were prohibitively expensive. They also had to be seismically and acoustically quiet, have such amenities as telecommunications and nearby housing readily available, and be set at least 1,500 miles apart to eliminate signals generated by local noises. As with Weber's setup, when the vibrations from a passing train, truck, or other jostling event show up in only one of the detectors but not the other, it can be ignored.

Operating in unison, the two facilities vastly broaden the search for gravity waves. Bars when they were operating could theoretically detect a supernova going off in our galaxy, but the chances of that occurring were

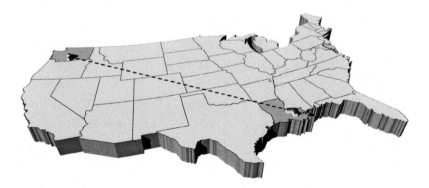

To rule out local disturbances, the LIGO sites are separated by nearly 1,900 miles.

always slim, once every thirty to a hundred years. Two neutron stars colliding somewhere in the Milky Way occurs only once every hundred thousand
years. That's not good odds for a steady career. LIGO was sold on the idea
that it would expand the search to the myriad galaxies beyond our galactic
borders, giving researchers a better chance of regularly seeing gravity-wave
events. More than that, interferometers provide far more astrophysical
information about a source than a bar.

The structures in Washington and Louisiana are very similar, though at
first they were more like fraternal than identical twins. The Hanford facility
for a while actually housed two interferometers, operating side by side
through the arms. There is, of course, the full-length interferometer of four
kilometers, but there was also one half as long. New equipment originally
destined for the shorter version is now being donated to India for that
country's planned gravitational-wave observatory; LIGO scientists agreed
that a third detector situated elsewhere was far better for the science.

The LIGO detectors in the West and South are now receptive to frequencies from around 30 to 3,000 hertz (with the aim to get down to 10 hertz).
On the musical scale that roughly extends from the lowest B note to the
highest F-sharp on a piano, almost its entire range. Within that broad band
observers are beginning to look for all kinds of signals from the cosmos.
Given the audio range of the signal, there might be single cymbal crashes
from exploding stars, the continuous hum from a swiftly rotating pulsar, an
extended glissando (a rapid ride up the scale) from the merger of two black
holes, as well as a faint background hiss, the gravitational equivalent of the
cosmic microwave background (more on gravity-wave events in the chapter
"The Music of the Spheres"). Moreover, LIGO is set up to compare its findings with other heavenly data gatherers, such as other laser interferometers,
astronomical telescopes, and neutrino detectors, just in case another visible
and spectacular event, such as Supernova 1987A, occurs on its watch.

A gravity wave washing up on the shores of Earth affects both detectors
nearly simultaneously. But in the process the interferometers face an array
of interferences that Weber never had to worry about, which makes this
endeavor seem all the more astounding. Take ocean waves, for example.
Each LIGO detector is located miles from a sea, but that doesn't prevent the
planet's bodies of water from introducing a noise. When waves hit the

shores of North America, all over the edges of the continent, they collectively produce a low reverberation roughly every six seconds, a microseis-
mic growl that peaks at 0.16 hertz. That's a low note to best all low notes.
In fact, it's so low that the vibration travels through the Earth with ease. "If
you just have a mirror sitting there," said LIGO physicist Michael Zucker,
"it feels that rumble." So LIGO's mirrors get a tiny push. "It's a headache,
a huge headache during Louisiana hurricane season," adds Zucker. LIGO
is also affected by Earth tides, deformations of the Earth caused by the gravitational pull of both the Sun and the Moon. The effect is slight—several
millionths of an inch—but still noticeable enough that LIGO has to take it
into account. Tidal actuators cyclically push and pull on the optical tables to
compensate.

More worrisome, perhaps, are thunderstorms within the vast heartland
of the United States. It's one of the few outside sources of interference that
has a chance of being felt simultaneously by both widely separated detectors. When a thunderstorm, say in Utah, unleashes a lightning bolt with
millions of amps of current, that bolt produces a magnetic field that propagates over the entire country. It's possible that both sites will see it at the
same time. Such a coincident signal could be misinterpreted as a gravity-
wave event. To reject these imposters, magnetic coils have been set up at
each site to monitor these magnetic pulses. That's why no CB radios are
allowed on the sites as well, since they might also interfere.

LIGO has evolved into an ever-growing enterprise. The construction cost
of the first-generation LIGO, which began operation in 2001, was $292 mil-
lion, plus more for commissioning and upgrades. An additional allotment
of $205 million was later allocated for the second generation, what is called
Advanced LIGO. Further funds to operate the LIGO laboratories and other
small NSF-supported groups in the United States brought the total over
some forty years to around $1 billion. That made it the single most expen-
sive undertaking ever funded by the National Science Foundation. (The
doomed Superconducting Super Collider, which was to cost some $8 billion
or more, was mostly funded by the Department of Energy.) LIGO's in-house
staff now numbers some two hundred workers at Caltech, MIT, and the two
detector sites. In addition, there are about eight hundred researchers at
other universities and scientific institutes throughout the world who are

also funded for ongoing work on data analysis and technology improvements. Together, they comprise the LIGO Scientific Collaboration.

Stan Whitcomb saw this evolution from tabletop science to large-scale venture firsthand. Having previously worked on instrumentation for submillimeter radio astronomy telescopes, he came to Caltech in 1980 to help assemble the forty-meter prototype imagined by Drever. Within five years, though, Whitcomb left for industry, in part because of his discouragement over the slow progress. "In 1980 we had assumed that we'd have large detectors in place within eight years. But we hadn't recognized all the technical problems. We didn't appreciate how difficult it would be to make the sensitivity improvements. We might not have gone forward if we knew ahead of time," he said. "We [had] six chunks of glass—the four test masses, a beam splitter, and a recycling mirror—all suspended on wires. We [couldn't] touch them, yet we still [had] to position them to within a billionth to one hundred-trillionth of a meter." (Advanced LIGO requires even more precision.) Whitcomb returned to LIGO in 1991 when he saw the NSF make its stronger commitment to the project, and he continued working on it until his recent retirement. Advances in technology made him more optimistic. "We were blissfully ignorant and lucky that technologies . . . advanced to help us—advances in supermirrors, lasers, and vacuum systems," he said. Each piece of the detector is a marvel of engineering.

LIGO will always be a work in progress. Before any upgrade comes a plethora of engineering decisions, each choice affecting another in a seemingly endless stream. What kept LIGO researchers on course amid these storms of detail? "People took pleasure in solving these technical challenges," answered Peter Saulson, "much the way medieval cathedral builders continued working knowing they might not see the finished church. If there hadn't been a fighting chance to see a gravity wave during my career, I wouldn't have been in this field. It's not just Nobel fever. Maybe it was a risky choice when I was just coming out of graduate school, but now I know it was a good decision. The levels of precision we strive for mark our business; if you do this, you have 'the right stuff.' "

When new optical telescopes come online, such as the twin Keck telescopes in Hawaii or the Hubble Space telescope, there is usually a celebratory "first light" event, the moment when the instrumentation is turned on and the first picture taken. LIGO's initiation was not so dramatic. Because of the complexity of its engineering and optics, initial LIGO required a few years for its initial shakedown before it reached the point when the two interferometers—the one at Hanford, the other in Louisiana—could work in concert with each other for a considerable fraction of each day. Then and only then did the search for gravity waves really begin.

The first-generation detectors were not expected to register a thing—and they didn't. "We're amateurs in a way," said Weiss in LIGO's early days. "We just hope we've made all the right decisions." Weiss never worried like that with the COBE satellite because COBE was an extension of past measurements. Everyone knew in some way what to expect. Gravity-wave astronomy, by contrast, was brand-new territory. Its scientists had built instruments for an effect that had never before been detected directly. Failure, Weiss feared, would have stalled the field for a very long time. For its critics that made LIGO technologically unjustifiable and premature. However, LIGO was built on the knowledge that solutions could not have been obtained without first building a full-sized facility to carry out the needed preliminary tests (and the belief that those solutions would lead to the detection of gravitational waves). As noted earlier, there was a reason LIGO is called an "observatory." Its builders did not intend to carry out a single experiment but to operate the facility for decades to come, much the way the great two-hundred-inch telescope on California's Palomar Mountain has been used and upgraded since 1948.

The first LIGO was constructed with Weiss's original strategy in mind— build *big,* then study the instrument as it operated to improve and design the innovative equipment needed to reach the sensitivities that could detect cosmic sources. All the operations of LIGO in the early 2000s were carried out with that goal in mind. Even though initial LIGO was a hundredfold jump in size from previous laboratory detectors, it remained a testbed. It reached its highest sensitivity—a strain of around 10^{-22}—in 2007 and looked for signals until 2010. None ever registered.

Yet even as the initial detectors were installed and turned on, laboratories around the globe began work on LIGO's next-generation instrumentation. The LIGO Scientific Collaboration studied new materials, improved the lasers, bulked up the mirrors, and tested novel vibration isolation methods. "This was hard work," said LSU physicist Gabriela González, "invisible to most people outside the LIGO collaboration." By 2011 the two observatories were shut down to install the newly designed hardware, a four-year endeavor that was well worth the wait. Advanced LIGO was turned on in 2015, and within months of commissioning, it was three times more sensitive—able to register events three times farther out than the initial LIGO. And three times farther out in all directions meant that the total *volume* of space accessible to LIGO increased by three cubed, a factor of twenty-seven. Consequently, it had the potential to detect twenty-seven times more extragalactic events over a month of observing. Improvements in sensitivity continue as I write this. The ultimate aim is to improve Advanced LIGO's sensitivity over the original LIGO by a factor of ten. That translates into 1,000 times more volume in the universe to examine. "That gives us a factor of 1,000 for finding some extragalactic thing like a coalescing black hole. You go from one event every ten years [with initial LIGO], which is pretty painful, to an event every three days, which is very nice. Reasonably small gains are very important," noted Kenneth Libbrecht. But the overall performance depended on lowering all the various sources of noise together, not just one or two at a time. "It's like a limbo dance," suggested Libbrecht. "You [had] to lower the bar for all of them." And those components— mirrors, suspension, seismic dampeners, vacuum system, lasers, and interferometer controls—as you will see, are varied indeed.

Some of the most crucial elements of an interferometer are its mirrors. GariLynn Billingsley has been the guardian of LIGO's mirrors for more than two decades, first overseeing the smaller mirrors used for the first-generation LIGO, then afterward the bigger ones installed in the advanced upgrade. Working from Caltech, she has monitored each step of their production. "Their manufacture was a heroic effort," she said. These pieces of glass pushed the limits of mirror construction, because their specifications were well beyond normal industry standards. Since the end mirrors are parked four kilometers from the laser, they must reflect the light up and down those long corridors extremely accurately. The mirrors now used by Advanced LIGO ended up with a surface so smooth that it didn't vary by more than four billionths of an inch, especially in the central six inches (sixteen centimeters)

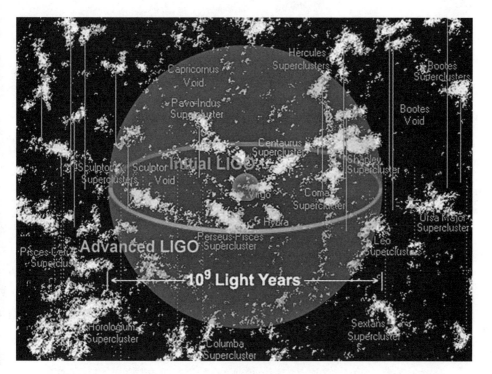

Initial LIGO could look for binary neutron star mergers within some fifty million light-years of Earth; once it reaches full sensitivity, Advanced LIGO will see such events out to distances of half a billion light-years. Being stronger, black hole collisions are already being seen more than a billion light-years away. *Beverly Berger, NSF, using a galaxy map by R. Powell (http://www.atlasoftheuniverse.com).*

where the laser beam is aimed. "Imagine looking across the diameter of the Earth, if it were that smooth. Then the average mountain wouldn't rise more than a third of an inch," pointed out Billingsley.

Mirror construction started with choosing a material that reduces a major source of interference in an interferometer: thermal noise. At room temperature the atoms within the mirror are continually vibrating, movements that could easily mask an incoming gravity wave. But if the mirror material acts like a bell (the engineering term is having a "high Q" for quality factor), those jiggles are confined to certain narrow frequencies. Restricting the noise to those specific bands allows other frequency windows to remain open for gravity-wave searches, free of interfering noise. It's akin to pushing the furniture in a room off to one side, leaving the remaining space clear for use.

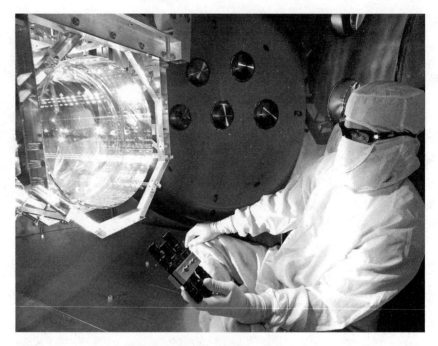

Angling a light on its surface, a LIGO technician inspects a mirror (test mass),
suspended within the mirror's active isolation system. *Courtesy of Matt Heintze/
Caltech/MIT/LIGO Laboratory.*

Fused silica was the material of choice. It is a type of glass that can be
manufactured exquisitely pure and uniform. For the advanced test-mass
mirrors, this was supplied by Heraeus in Germany. But that was just the
first leg in a worldwide journey. The glass was then sent to Zygo Extreme
Precision Optics in California for polishing. What resulted were cylindrical
disks about thirteen inches (thirty-four centimeters) in diameter and eight
inches (twenty centimeters) thick. Each weighs eighty-eight pounds (forty
kilograms). The final step was applying the reflective coating, a job performed by Laboratoire Matériaux Avancés in France. The thin covering is
composed of alternating layers of silicon dioxide and tantalum pentoxide.
For every one million photons of light hitting the mirror, only four are lost
with each bounce off this surface. And for every ten million photons, only
two are absorbed. Such exquisite reflectivity maintains the laser's power
and reduces heating, which keeps the mirrors from degrading.

Twenty mirrors were ultimately made. Each interferometer uses four of these mirrors as test masses. The rest are spares. In addition, an assortment of secondary optical pieces was polished and coated to similar standards for use in aiming and guiding the laser beam in the interferometer.

The spares are kept in the LIGO optics laboratory, situated in the bowels of the Downs Building on the south end of the Caltech campus. The disks are stored in what look like aluminum cake covers set on rolling shelves. The lab itself, a windowless room, resembles a hospital operating room. Visitors and technicians wear masks, caps, and booties at all times to keep the environment clean—even a short breath on a mirror would seriously contaminate its surface.

Looking down upon a glass disk is like looking into a pond of pure, perfectly still water. To the naked eye there is nary a bubble or scratch. All LIGO optics team members are extra cautious when handling a mirror (with the help of a robotic arm), and they have good reason to be nervous. Each mirror costs $250,000 to $400,000 and a year or more to manufacture from start to finish. "We've never dropped an optic," notes Billingsley, "but the fear runs strong in us!"

Silica also proved a decisive material for suspending the mirrors. Within the observatories, each was hung from what looks like an ultramodern gallows. The mirror is suspended—hangs freely—as the end of a four-stage pendulum. If any noise starts traveling down the wires, each stage dampens it more and more, until the test mass feels no other force except gravity. The entire system separates the mirror (the test mass) from the rest of the equipment.

In the initial LIGO each test mass was balanced on the slimmest of supports: a steel wire as thin as dental floss. The material was chosen to keep both the tone as pure as possible and the string vibrating as long as possible. It's like a having a concert where the single tone of a violin lingers for several minutes after the music is over. In this way a gravity wave with a lower or higher frequency can be better distinguished, not hidden within a cacophony of thermal noise. But even before LIGO was in operation, they knew that the metal wire would exhibit lots of thermal noise from 40 to 100 hertz, a crucial frequency range for their target sources. This led to the invention of silica suspension wires for Advanced LIGO. Fused to

LIGO team readying and installing one of Advanced LIGO's new optic and suspension systems. *Courtesy of Caltech/MIT/ LIGO Laboratory.*

each side of a mirror, these slim glass fibers (less than a millimeter wide) lowered thermal noises in the low-frequency range by a factor of one hundred. If used as violin strings, they would ring for a *month* when plucked.

Less exotic than thermal noise, more commonplace vibrations also have to be monitored and eliminated. Gravity waves cause the distance between the four-kilometer-separated mirrors to change quickly; the space measured between the mirrors rapidly jitters. Advanced LIGO senses the movements that range from thirty to three thousand times each second (30 to 3,000 hertz). On the other hand, the movements induced by normal geologic processes are relatively "slow." They can cause a mirror to oscillate just one time a second. The motion is usually quite tiny, the width of a human hair. But if the mirror should do this, move back and forth over one second (a 1-hertz vibration), observers can simply ignore the motion, essentially filter it out. It's one of the tricks that enables researchers to zero in on the incredibly tiny space-time movements introduced by a gravity wave. Tidal movements are so sluggish, compared to a gravity-wave signal, that they don't mask the cosmic signal at all.

But LIGO must be isolated from many other terrestrial motions: a work truck driving by, a high wind hitting the building, or a seismic tremor (both small quakes nearby, such as the ones in the Midwest caused by fracking, and particularly strong earthquakes from around the world). Vibrations from these activities might still introduce a noise that either mimics a gravity wave or interrupts the interferometer's operation. The earliest line of defense is the floor itself. It is a slab of concrete, thirty inches (seventy-six centimeters) thick, which first dampens vibrations coming from the outside. That has not changed since LIGO was constructed. As a second line of defense, initial LIGO's test masses and other optical equipment had their own seismic isolation systems. Each test mass was suspended beneath a series of stainless steel disks, with each level separated by a set of elastic springs. This arrangement highly damped any ground motions, much like the suspension in a car. It cut down the seismic noise by a factor of a million, but even that was not enough to carry out good observations. "When it got shaken up," said David Shoemaker, director of MIT's LIGO laboratory, "it would ring on for a period of time, making it difficult to get the machine back in working condition. We fought this daily. There was a train in Louisiana that passed at regular times, knocking the system out of lock."

That frustration led to one of Advanced LIGO's key design changes. Each test mass platform is now hung from an *active* isolation system. No longer are outside vibrations merely suppressed. Sensors mounted in this new isolation system directly detect any movements and rapidly activate motors to counter the movements. It eliminates the incoming motion appreciably. This keeps each test mass, hanging below, as steady as a rock. This improvement, coupled with the glass wire suspension, added two octaves to the low-frequency range of LIGO, a major requirement in searching for black hole collisions. The train at Livingston is no longer a problem, although earthquakes from afar can still get them "out of lock" (that is, break the required sync of their instrumentation) a few times a week. When that happens, "we're prepared to go get coffee and some reading material," said Caltech postdoc Anamaria Effler, because it takes an hour or two for the Earth to stop ringing from the tremor.

The largest portion of the funds for initial LIGO, about 80 percent, did not go into sophisticated equipment, such as electronics or computers. Rather, it went into the low-tech items of this high-tech enterprise: constructing the pipes, pouring the concrete, and building the vacuum pumps. All stayed intact for Advanced LIGO. The stainless steel tubes were fabricated by a method regularly used to make oil pipelines. Temporary factories were set up near both LIGO sites to carry out the task. In Louisiana the beam tubes were constructed in a vast warehouse next to a shopping center some twenty miles from the observatory. Over eleven straight months, workers from the Chicago Bridge and Iron Company produced four hundred separate pipes, each sixty-five feet (almost twenty meters) long. The stainless steel arrived at the warehouse on immense rolls, like giant-sized rolls of kitchen foil. On an automated conveyor, a roll unwound, with the continuous sheet sent into a machine that coiled it helically and then welded it automatically with a high-frequency pulse of energy. Rolling along at forty inches per minute, a sixty-five-foot-long section could be completed in about forty-five minutes. The finished tube shimmered with stripes of dull red, blue, and cream, resembling a designer barber pole. Each pipe was individually tested for leaks (not one failed) and then taken to its final cleanup.

In a separate room a small robotic cart was sent into each tube, which washed and steam rinsed the entire interior. In the end what remained was one part dirt for every million parts of water. "Each detail [was] like this," said LIGO engineer Cecil Franklin. "And any one error [could have caused] problems down the line." Each tube was finally encased in a plastic bag—what looked like a body bag—for transport to the site, where the separate tubes were finally welded in line to form the long arms. All in all a complete interferometer has thirty miles (forty-eight kilometers) of welds.

The ends of each arm are actually situated several feet higher off the ground than their starting point at the center station. That's to compensate for the Earth's curvature. The gradual rise keeps the laser beam tubes positively straight as the Earth curves downward. With the Hubble Space Telescope debacle on his mind, where a small measuring error in the factory led to an out-of-focus mirror, Weiss was waking up at night in a sweat during LIGO's construction wondering if the tubes were truly arrow straight. Then he realized that they could just look down a completed arm and check.

"We got down at the end of one tube in Livingston and asked someone to hold a big searchlight at the other end," recalled Weiss. "We could talk to them because the tube is a good acoustic waveguide. The light was turned on, and when we looked it was about thirty centimeters low. We called out, 'Why don't you put it in the center?' They replied, 'It is in the center.' Within fifteen seconds, it dawned on me what was happening: bending of light by the atmosphere, due to both the change in air density from the top of the tube to the bottom as well as temperature differences. The laugh is we were lucky. It was early morning, before the Sun really beat down on the apparatus. If we had waited just three hours, we wouldn't have seen any light at all. I would have assumed we had left something in the tube and demanded to go back in, which would have been expensive."

Once a laser beam is pointed down a tube, air is its biggest obstacle, which is why the arms are evacuated down to a trillionth normal atmospheric pressure. That's to prevent light from scattering off stray gas molecules and introducing a noise that might look like a gravity wave. With each detector taking up three hundred thousand cubic feet (8,495 cubic meters) of space, LIGO ended up creating one of the largest artificial vacuums in the world (only surpassed by the Large Hadron Collider in Switzerland and a NASA vacuum chamber in Ohio). The atoms that managed to remain in the pipes would fill only a thimble under normal atmospheric pressure. LIGO accomplished this feat by not following conventional wisdom. Instead, it took a risk on a radically new procedure. LIGO researchers arranged for a special steel, an alloy that was cooked for several days to remove excess hydrogen to levels a hundred times less than those of commercial vacuum systems. Ordinarily, hydrogen atoms leak out of steel, which can clog up a vacuum. Designers of particle accelerators deal with this problem by heating their pipes after assembly. This excites the hydrogen molecules enough to coax them out of the metal, where they can be vacuumed away. But such pumping would have been a terribly expensive process for pipes four feet wide and miles long. Limiting the hydrogen in the steel from the start made the price of LIGO affordable. Though the pipes were still heated for thirty days to eliminate the last remaining gases, the number of pumps required to suck away those stray molecules was sharply reduced. "Otherwise, LIGO would have been far too costly to construct," said Whitcomb.

That vacuum, first achieved in 1997 and 1998 at Hanford and Livingston, respectively, has remained uninterrupted within the tubes to this day, because it would be too expensive and time consuming (possibly a year or more) to break the entire vacuum, carry out maintenance, and pump the air out of the tubes all over again. (Localized vacuums are broken to make repairs or insert new equipment.) This doesn't mean that the vacuum system hasn't faced problems. The Livingston observatory, at one point, was experiencing tiny leaks in one of its tubes. Hunting down the leaks' locations along the many kilometers took great ingenuity. After months of searching, researchers discovered that the tube's insulation, left over from its bake-out, had become the home of mice and insects, whose waste products were corrosive to the stainless steel. A wholesale cleaning of the tube exterior, plus patching the microleaks with epoxy, stemmed the problem, at least for now. "We're like the Dutch boy with our fingers in the dike, but we have only so many fingers and toes," said Zucker. From here on out, added Shoemaker, "it's going to be a major focus of the LIGO laboratories to maintain the infrastructure."

The LIGO infrastructure was built to have a twenty-year lifetime, a period that is almost up, even though its astronomy searches are just beginning. So, LIGO researchers are looking into schemes that can either keep vacuum leaks at bay or make a maintenance shutdown last far shorter. "We want to use LIGO another twenty years," stresses Shoemaker. With all its ultramodern and mind-bending equipment, LIGO wouldn't work at all without this basic, old-fashioned need—sustaining a vacuum.

As Ron Drever discovered early in his investigations, a gravity-wave interferometer also requires a laser that is amazingly stable. Any change in frequency or intensity might be mistaken for a gravity-wave effect. Originally, LIGO was designed to use an argon ion laser, a type of laser that emits a brilliant green light. "But having the argon ion laser was like using a radio with vacuum tubes after the development of transistors," said Barry Barish. At the last minute they switched to a solid-state infrared laser, which is extremely stable. Small and powerful, they are called neodymium YAG (for the yttrium-aluminum-garnet crystal that lases). Over a billion trillion cycles of light, its frequency doesn't vary by more than one cycle. Moreover, this laser early on showed promise in scaling up its power. Initially, LIGO used a laser with 10–15 watts of power, but Advanced LIGO now uses a version able to amplify that to 35 watts, and plans are under way to boost that to 200 watts, which would help reduce an interference known as shot noise. As Einstein first noted in his Nobel Prize–winning discovery, light travels in discrete bundles called photons. When those particles of light hitting a mirror are few, the count is rather noisy. Consider the noise emanating from a slow-dripping bathroom sink. The sound of the individual drops is easily noticeable (and quite annoying) when the flow is small. But the noise smoothes out and gets quieter when the flow increases to a steady stream. Similarly, when the laser power is increased in an interferometer, the relative strength of its shot noise is lessened. Recycling the light also decreases shot noise. By increasing the laser power in these ways, LIGO is able to "see" much farther out into the cosmos. A hefty improvement in laser power—and the concomitant decrease in shot noise—allows the detector to measure smaller space-time strains arriving from events far more distant.

Together, these varied pieces of equipment form the interferometer itself, the very heart of LIGO. No interferometer, no discoveries. Zucker at MIT, who cut his teeth on the Caltech prototype as a graduate student in the 1980s, headed up the initial LIGO task group on interferometer control. "It's the glue that holds the optics in alignment, to make sure that the lasers are resonating in the proper way," he said. The strategy is to have the photons circulate as long as possible (which improves sensitivity). Based on the reflectivity of the latest mirrors, they expect about 280 bounces. In each journey down an arm and back, the waves must march lockstep in synchrony with one another. They must stay in phase. To do this, exquisite timing is demanded. The software must know exactly what time it is, so that every element of the system can stay in synch. Like a human heartbeat skipping a beat, losing track of time would lead to trouble. LIGO consulted the field of control system engineering to develop its elaborate feedback system. "But we have a pretty unique situation," said Zucker. "Most control system engineers' jaws dropped when they heard what we were trying to do."

When the light waves are getting out of step because a mirror has moved, a force from the active seismic isolation system is applied to keep the arm lengths equal. In other words, when a gravity wave (or other noise) pushes on a mirror, the controls are programmed to pull it back into place. This includes a final stage that stabilizes the test mass mirrors themselves. An electrostatic drive, positioned close to but not touching a test mass, produces an electric field that penetrates the mass. An extremely small force—similar to the one between a charged balloon and a wall to which it is attracted—can then be applied to the mass by varying this field. Photon pressure from a tiny laser is used to check the sensitivity of the entire system. The gravity-wave signal is actually hidden in all these manipulations. By keeping track of the forces needed to keep the arm and test mass firmly fixed, the interferometer is in essence recording the gravity wave itself. The movements needed to counter the gravity wave's force echo the strength and frequency of the wave itself.

But once you have that wave in hand, how do you prove that it's truly a gravity wave? How do you recognize a gravity wave signature, when one had never been seen directly before 2015? In particle physics experiments, physicists are often searching for specific, discrete events. Particles collide

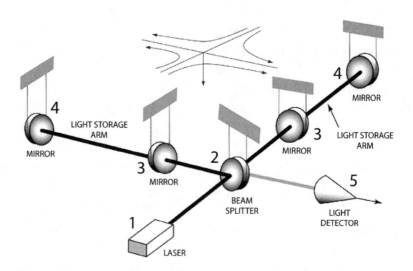

The beam of an infrared laser (1) enters the interferometer and passes through a beam splitter (2) to form two beams. Each beam is directed down an arm and reflected many times between a central mirror (3) and an end mirror (4). Eventually, the beams exit the arms and are recombined. If a gravity wave passes through, altering the length of the arms, a light detector (5) detects the change in the resulting light pattern of the recombined beams. *Courtesy of ESA.*

at a fateful moment in an accelerator, creating a burst of energy that instantly transforms into a plethora of new particles. The debris is there to sift through. The long history of particle physics has given physicists a good feel for distinguishing a real event from a fake.

In gravitational wave physics, in contrast, researchers are dealing with a continuous stream of unknown data. In these early days of the field, LIGO depends on a combination of simulations and equipment monitoring to point the way to the science. They track down all the known sources of interference in the interferometer, such as the seismic, thermal, and shot noises. LIGO researchers are like doctors who learn and characterize every symptom of their patient. This helps them locate and correct initial sources of noise, be it a misaligned mirror or a noisy cable line. "There are diagnostic probes all over the instrument," noted Hiro Yamamoto, who initially directed the simulation efforts. Over time they have come to understand each detector's unique personality and maintain a catalog—a set of red flags—characterizing its typical noises. In this way, a gravity wave stands out as something different. For example, at the Hanford observatory a series of glitches around 60 hertz once appeared in one of the channels about every seventy-four minutes. These events seemed to be correlated with some magnetometers on site. This clue led them to the culprit—a refrigerator compressor! The noise disappeared once the refrigerator was turned off.

Yamamoto first worked on simulations for the Superconducting Super Collider. "It took me some time to adjust my common sense, the way to understand an event in gravity-wave physics," he said. But there is some common ground: in both particle physics experiments and gravity-wave searches, scientists have to understand the instrument down to the smallest nut and bolt to be able to distinguish the background noise from the signal.

Sensors are installed throughout LIGO, all providing data on the detector's condition as well as any signal it might be receiving. These sensors keep track of the laser noise (changes in the light's intensity and frequency, which could mimic a wave), electromagnetic interferences, and any geophysical or terrestrial phenomena that might wiggle the test masses, such as seismic tremors or particularly loud acoustic noises. Data stream in from microphones, seismometers, tiltmeters, magnetometers, power-line monitors, weather stations, and cosmic-ray shower detectors. (Normal office noise does not affect the system.) These data are collected at each observatory in some two hundred thousand separate channels simultaneously, tracking an event like a movie. Each frame is a snapshot of an interval in time—the signal plus all the attendant instrument and environmental noises. "The computer must do a cycle in real time every sixteen microseconds and never miss—ever, ever, ever, ever," stresses Keith Thorne, Livingston's software engineer. To carry this out, they use the same hardware as high-frequency traders on the stock market.

At each observatory this information continually flows at some thirty million bytes (megabytes) per second, twenty-four hours a day when in full observing operation. With around thirty-one million seconds in a year, each interferometer gathers about one petabyte of data (one thousand trillion bytes) annually. The data are shuttled in real time to banks of computers at various locations, where the bytes are filtered and stored. The data are kept in a standardized format, a procedure that is also being used at other gravity-wave observatories around the world. This allows each observatory to exchange and compare data, which is vital for confirming a potential signal. About a third of these data are eventually stored permanently at a Caltech archive on disks and tapes, while the rest are kept temporarily until no longer needed.

The movements of the mirrors themselves, monitored by the laser light, are recorded on a special channel—the gravity-wave channel. It accounts for less than 1 percent of the total data collected, but if a gravity wave comes by, that is the channel where it is found. Unlike with an optical telescope, no pretty picture of the source is obtained. Visible light waves are quite small compared to their sources, be they gas clouds, stars, or galaxies. Such waves can hit a receiver, say a photographic plate, and produce an image of the celestial object. Gravity waves, instead, are often as large or even larger than their source. A gravity wave with a frequency of 1,000 hertz, for example, spans nearly two hundred miles from peak to peak. Such a signal resides in the same frequency band as audio. You can actually listen to the signal once it is electronically recorded, just as Robert Forward did with his early Malibu detector.

Computers, not ears, though, sift through LIGO's data. Picking out a definitive signal from a chaotic profusion of bits and bytes is not a totally new endeavor. Though difficult and challenging, it's similar to the way in which military sonar experts search for the distinctive sound of a submarine amid the many noises of the sea. Essentially, as the data stream comes in, it is compared to a template, a theoretical guess at what a gravity-wave signal might look like. Take, for example, the case of two neutron stars spiraling into each other. Of course, the exact nature of the gravity waves being emitted from such a system will depend on both the masses of the neutron stars and their orientation as viewed from Earth. So there are many possible wave patterns. Teams of numerical relativists and theorists have been working for years on generating the myriad variations and having them encoded into LIGO's search software. When it is in observing mode, LIGO continuously compares its stream of data against a few hundreds of thousands of signal patterns throughout the day and night, each pattern representing the waves emitted by differing types of events. Fortunately, computers have now achieved speeds that can handle such a load.

In real time, both sites are on the lookout for certain classes of events that don't get repeated—a star exploding (called a burst) or binary stars colliding (a binary coalescence), for example—by continually comparing for timing, similarity of waveform, and local interferences (to reject spurious coincidences). Once a candidate pops up, it is immediately compared with the environmental and instrument channels to see if it was just terrestrial noise. If the candidate appears genuine, the astronomical community is quickly notified of its time and general location in the sky to maximize the possibility of catching any electromagnetic radiation emanating from the transient event.

But all the observing data that are collected and recorded are always reexamined offline at some point. Called a deep search, this reappraisal may catch weaker transients missed by the real-time searches, but it is also used to hunt for continuous sources, say from a deformed neutron star that stirs up space-time as it rotates. Those gravitational waves would not emerge out of the noise until data are gathered over long periods of time.

In the 1990s, as LIGO was under construction, most researchers expected that it would take a long and grueling procedure to judge a gravity-wave candidate. They envisioned that the signal would be weak, buried within a sea of noise, and requiring months to eliminate all the possible sources of interference. There would be myriad arguments before consensus was finally reached. If that were the case, when would you really have a first detection?

Weiss decided to set up ground rules. At a meeting of gravity-wave researchers at the Livingston site in 1998, he threw down the gauntlet. "The heart of the matter," said Weiss, "will be detector confidence." First and foremost, the observation had to occur in both LIGO interferometers effectively simultaneously. Specifically, there had to be a believable delay (no more than ten milliseconds of travel time) between Hanford and Livingston, as well as no outstanding environmental interference. Both sites had to see the same spectra, the same frequency, and the same amplitude. "We're spending a lot of money, so it's crucial to be careful," he stressed.

Weiss offered the following strategy: "A detection of gravitational waves is to be announced only after a statistically meaningful analysis has been performed of the data of ALL instruments that were observing. . . . The data and statistical results are brought to a council composed of representatives from each observing group. The initial publication is submitted in two parts. A paper from the group(s) making the observation and their analysis and a second paper from the council discussing the statistical significance in regards to the worldwide effort, in particular, the probability and confidence of detection in some of the instruments as well as the reasons for non-detection in others." What if, posed Weiss, one group announces and others disagree? "Then," he concluded, "we'd have gone full circle to Joe," referring to the disagreement that persisted between Weber and the rest of the gravitational-wave community over what his bars were picking up.

Fortunately, Weiss was prescient, and his plan was followed almost to the letter when first detection occurred at LIGO. But his critics were right, too. At that 1998 Livingston meeting, veterans from the high-energy physics community warned Weiss that leaks to the press about a possible detection would likely add pressure. What they all didn't foresee was the rise of the internet and social media in the intervening years. It turned out that LIGO leaders had to face a cascade of gossip on Twitter in the months between the detection and its official announcement.

"We're attempting to find a jewel within a dense forest," Yamamoto had noted in the 1990s. But that celestial forest was richer than anyone had expected, allowing the jewel to stand out like a glowing treasure. First detection in the fall of 2015 occurred with a "golden event"—the fairy-tale signal that LIGO scientists never believed would truly arrive first . . . but did.

The Chirp

BY MARCH 2014, LIGO RESEARCHERS HAD AT LAST completed the installation of their advanced equipment at both observatories and afterward began a long process of commissioning to test, tune, calibrate, and make sure every piece was working up to spec. In LIGO's early years, Livingston always had more seismic problems than Hanford, so spirits soared when the Louisiana site was able to lock its interferometer (keep it in full operation) reliably for longer than twenty-four hours. More than that, each observatory quickly ramped up to a sensitivity that was more than three times better than the initial LIGO had ever achieved.

By August 2015 the LIGO teams began preparing in earnest for their eighth engineering run, the last check before they launched their first observing stint, known as "O1." No one fully expected a detection during Advanced LIGO's maiden voyage, but in preparation for the upcoming observations many of the engineers, scientists, and technicians worked around the clock in hope of beating the odds.

There were still problems to wrestle with. In mid-August, Rai Weiss traveled down to Livingston to assess a radio interference problem. In the push to get Advanced LIGO built and operating, many transformers and cables had not been properly shielded, which limited the interferometer's radio-frequency sensing systems and could possibly spoil any measurements. The same was true at Hanford. By the first week of September, Weiss was recommending that LIGO postpone its observing run by one or two weeks to fix the problem. But others vetoed that suggestion, not wanting to lose the momentum toward O1. Many researchers were already locked into travel plans to get to the sites, and the problem hadn't yet restricted LIGO's sensitivity.

On Thursday, September 10, the planning committee met to officially set the start of the observing run, expecting observations to commence in four days, that coming Monday. Both the detector and commissioning teams reported a "go" to proceed. But the calibration and data analysis groups told the committee that they needed more time for final tests and last-minute installations before O1 could start. It was a disappointing turn of events, but not unusual in such physics endeavors. "Not a big deal, just a few days," thought MIT physicist Lisa Barsotti, chair of the planning committee. "It's not like we are going to detect gravitational waves on day one."

So, in the wee hours of September 14, measurements and calibrations continued at both observatories. This was still an engineering run, but whenever the detectors were in lock, computers were automatically collecting and assessing data, as they had been for a few days. At 4:05 a.m. Central Daylight Time, William Parker, the Livingston detector's night-shift operator, entered a brief note into his electronic logbook. The observatory's current range was sixty-eight Megaparsecs (around 220 million light-years) for binary neutron-star collisions, and no seismic events were headed their way. The weather was clear. Parker was sitting in the control room, surrounded by about two dozen flat-screen monitors mounted along three walls. From there he could observe a variety of conditions: the interferometer's status, power levels, vacuum leaks, earthquake jiggles, and even the noise of human activity around the site.

LIGO scientist Joseph Betzwieser and University of Mississippi physicist Shivaraj Kandhasamy had been working late at Livingston to measure the response of the interferometer to the photon calibrator, which pushes the test masses with photon pressure. Such work kept the detector out of observing mode. On other days, they had often worked until daylight. But right about the time that Parker was typing into his logbook, the two decided to call it a night. "I could have kept going," said Betzwieser, "but I told myself, 'I need some sleep because I'm starting to make a lot of mistakes.'" They went home.

By then only two other people remained at the observatory with Parker. For weeks Caltech postdoc Anamaria Effler and Oregon physicist Robert

Schofield had been carrying out environmental noise tests, first at Hanford then at Livingston. "We take a speaker and blast noise, to see if it comes into the gravitational channel," explained Effler. "We have a paranoid approach— an anything-can-happen attitude. For magnetic fields, we have a giant coil and blast out magnetic fields." They were planning to finish up that night with a car injection—that is, "drive a big car next to the buildings, applying the brakes violently every five seconds exactly, so we can see if we can extract the pattern," said Effler. Such a test wouldn't have hidden an incoming gravity wave, but it would have compromised the data, flagging the information as unreliable. Unable to connect to the GPS satellites for timing, however, they decided to quit and drove off about 4:35 a.m. Only Parker remained at Livingston. The operator at Hanford, Nutsinee Kijbunchoo, also kept a lone vigil.

Fifteen minutes later, something happened.

At precisely 4:50 and 45 seconds, the Livingston detector registered a sizable event. Parker was unaware because the sound alert in the control room had not yet been activated. The same was true at Hanford, which silently recorded the signal seven thousandths of a second later. The data from both sites simply streamed into the computer pipeline.

There they were analyzed by the coherent waveburst (cWB) software, which is designed to use the two interferometer outputs coherently and find sudden spurts of excess power, like those from a supernova explosion. This software was operating during the engineering run, and when it saw something, another computer program was triggered that assessed the data's quality. With high confidence the data-quality software tagged the event as a glitch (too loud), but processing continued in any case. Within three minutes, an automatic message was sent out to the seven members of the coherent waveburst group, informing them of the event.

Marco Drago, an Italian LIGO collaborator working at the Max Planck Institute for Gravitational Physics (Albert Einstein Institute) in Hannover, Germany, saw the notice on his computer screen. It was nearly noon there. Just looking at a plot of the raw data—a moonlike crescent that depicted a brief surge that swiftly rose in frequency—the young postdoc knew immediately that, if it were a gravity-wave signal, it wasn't a supernova blast but rather the merger of two objects. The wave's frequency races up as the celestial bodies spiral into each other. But this chirp was so strong and clear (beautiful, actually) that he simply assumed it was either the insertion of an artificial waveform to test specific hardware during the engineering run or a blind injection—that is, a simulated wave secretly inserted into the interferometer to test everyone's capabilities in handling and deciphering a signal. This was nearly everyone's thought when first viewing the data—that's because such incidents had happened before.

In 2010, for example, toward the end of initial LIGO's operation, both observatories recorded what appeared to be a gravitational wave. Only a few people on the blind-injection committee knew whether it was a test, and their lips were sealed. The rest of the LIGO collaboration proceeded to analyze the signal as if it were a bona fide candidate. It, too, appeared to be the merger of two objects, and its strength suggested an event tens of millions of light-years distant. Because the wave appeared to be coming from the direction of the constellation Canis Major, everyone called it the "Big Dog" event. LIGO scientists spent six months weighing the evidence and preparing a scientific paper. In the end, the vote was unanimous. Everyone wanted to announce Big Dog as their inaugural detection. But first an envelope had to be opened (actually, a thumb drive file) to reveal whether a blind injection had instead occurred.

It turned out that Big Dog was a mock event after all, a bogus signal inserted by the blind injection team.

Was that happening again? Drago went down the stairs to the office of his colleague Andrew Lundgren and asked, "Do you know if they're performing an injection right now?" Lundgren checked the logs and said he was pretty sure they weren't, to which Drago replied, "Then, we have an interesting event." To be certain, Lundgren phoned the observatory control rooms, and Parker in Livingston picked up. "What is the current state of the detector?" asked Lundgren. "Is everything running nominally? Are there any injections being done?"

"We've got a good strong lock." answered Parker. "Everything is normal. There's no injections." At least, no authorized injections to test hardware.

Soon other people began gathering in Lundgren's office, with each asking whether it was an injection. When told no, someone remarked, "What do we do now?"

Drago decided to go back to his office and e-mail the burst group. "Hi all," he wrote. "cWB has put on gracedb [the *gravitational* candidate *event* database] a very interesting event in the last hour." He then provided his recipients the web links to the data. His list of recipients, though, included not only the small cWB group but also the master e-mail address for the *entire* LIGO-Virgo collaboration, more than a thousand members. "Not a good choice," Drago later reflected. His cWB colleague Sergey Klimenko at the University of Florida, who had seen the computer alert the same time as Drago, agreed: "After 8 AM [on the East Coast] people woke up, and the e-mail floodgates broke wide open."

"Slept through my alarm on the morning of the alert, haven't slept since," said Cody Messick from Pennsylvania State University.

His Penn State colleague Duncan Meacher chimed in, "First day back in work after defending PhD. So much for an easy start!"

"I was getting into the shower when an alert went off. . . . It literally caught me with my pants down," confessed Kipp Cannon, then at the University of Toronto.

Yet after years of never detecting a true wave, the LIGO collaboration had low expectations. They had a huge psychological hurdle to overcome. Nearly everyone's first response that morning became a mantra heard far and wide: "It must be a blind injection. It must be a blind injection." Over previous weeks, the four-member blind injection team had been preparing and calibrating the specific hardware that makes such a test possible. In fact, that's what Betzwieser and Kandhasamy had been working on before the detection. Inserting a blind injection requires delicately moving the mirrors using both laser photon and electrostatic forces to make it appear as if a gravity wave had hit. The larger community didn't know that a blind injection wasn't even doable as yet.

At Hanford, LIGO scientist Michael Landry, upon seeing Drago's message, hopped in his car and raced to the site to corner Jeffrey Kissel, a member of the blind-injection team. "I knew I could not ask him if he made a blind injection the night prior," said Landry, "but I could ask, 'Are we in a blind injection phase right now?'" Blind injections are never a total surprise; the LIGO community is made aware beforehand that a period is opening when it *might* occur. But Kissel replied, "No."

As reported in *LIGO* magazine, published by the science collaboration, Landry probed further: "Did you make a regular injection? Did you make a blind injection test on regular channel? Did you make any injection at all?" Each time, Kissel answered "no." Landry was astonished. Given this information, he confidently announced, during a 9:00 a.m. Pacific Daylight Time teleconference of the Detector Characterization group, that the recent event was *not* a blind injection. Those listening in from the Livingston observatory were stunned. "The reaction in the room . . . was very pronounced," said staff scientist Brian O'Reilly. There had been the matter of one data-quality program rejecting the signal and labeling it a glitch. But that computer-generated veto had already been removed (within two and a half hours of the detection) as soon as it was discovered that the channels involved had not been in an acceptable mode.

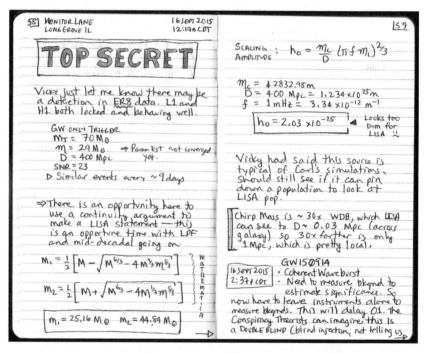

Soon after LIGO collaborator and Utah State physicist Shane Larson heard the news from Northwestern University physicist Vicky Kalogera, he started to make calculations in his work journal, just like many others in the LIGO collaboration.
Shane L. Larson.

From the teleconference, Landry went directly to Hanford's control room and initiated LIGO's prearranged checklist in response to gravitational-wave triggers. A parallel response was carried out at Livingston: a list was started of those who had access to the interferometers during the discovery, either in person or by remote, for future interrogation; data channels were checked to see if any physical or environmental issues could have generated a false signal. Over the ensuing weeks, everyone placed their ongoing findings into a special logbook on LIGO's website, where LIGO collaborators discussed the event among themselves in private (LIGO's standard observatory logbooks are open to the public).

This was not the way anyone imagined having first detection. First of all, the signal was so darn strong. "It was so large that we were all gobsmacked," said Keith Thorne.

"You don't expect a golden unicorn galloping across the screen," noted Mike Zucker. (The signal was intense for Advanced LIGO but would not have been strong enough for initial LIGO to have seen on its watch.)

And the wave arrived prematurely, before the LIGO researchers were fully ready to begin their scientific run: they had not yet collected a sufficiently ample background, data needed for assessing the validity of any gravitational-wave candidate. "We had no statistics on this brand-spanking-new detector's behavior," explained Joseph Giaime, head of the Livingston observatory. "If nature hands you the event right when you turn on the detector, you can't say anything about it until you've run for a while to find out how the detector behaves." They needed to know how often such a trigger might arise by sheer chance, perhaps from an electronic burp in the equipment.

That's why the decision was made on the fly to "freeze" both detectors immediately, changing none of their configurations in order to gather a background from each instrument. No electronics in each observatory were touched, no lights switched on or off. Work permits and truck deliveries were canceled to waylay any potential interference. "That plan was invented on the spot," said Giaime. With the run unable to shut down, the next day's regularly scheduled maintenance was rescinded. Rai Weiss, on vacation in Maine and unaware of the event, saw the cancellation notice on his computer and wondered if something had gone wrong or whether the LIGO staff was getting lazy. But David Shoemaker soon e-mailed him and dispelled those worries. Going onto the private log, Weiss saw a newly post-ed image and yelled out, "My god!" His son Ben, eating breakfast nearby, asked, "What's going on?" Breaking all the LIGO rules that gravity-wave candidates be kept hush-hush until announced publicly, Weiss decided to share the news with his wife and visiting family. "I couldn't keep it a secret in that house," he said. Coincidentally, Richard Isaacson, one of the NSF officials so crucial in keeping LIGO funded during its shaky, early days, was scheduled to join the vacationers. Once he arrived, he learned that his decades-old gamble might finally have paid off.

The image that had provoked Weiss's exclamation was not a graph of the initial signal but something else. Within hours of the detection, MIT physi-cist Matthew Evans had taken the raw signals captured at each observatory and passed them through a frequency filter, in order to strip out instrumen-tal noise and see what the waveform looked like over time in each detector's sweet spot—a window extending from about 30 to 350 hertz where the in-terferometer has the greatest sensitivity. It turned out that the wave stood out strong and clear. Its ripples depicted how space-time within the interfer-ometers stretched and squeezed multiple times as the gravity wave passed through. "It was amazing that anything could appear so cleanly in a plot like that," said Evans. It was because of the wave's extraordinary and surprising strength. Later that day Syracuse University physicist Stefan Ballmer carried out his own filtering scheme and time-shifted the Hanford observatory's waveform by seven milliseconds (the speed-of-light time difference between the sites for that particular signal), so that the Hanford wave overlapped with the one from Livingston. And even in this rough, not fully calibrated stage, the two signals matched beautifully—almost to every peak and trough. The graphs by Evans and Ballmer displayed the characteristics of the event that Klimenko in Florida had calculated soon after seeing the raw signal alone. Just eighty-five minutes after the detection, Klimenko had e-mailed his burst group that it was a "nice inspiral" involving two black holes weighing (from his first estimate) about twenty-seven solar masses. But now the approach and mighty collision were visually portrayed in Evans's and Ballmer's plots.

Black holes were a surprise. With certain knowledge that neutron star binaries exist, most LIGO observers figured that these compact stars would be their sure bet for first detection. (For his part, Kip Thorne was not

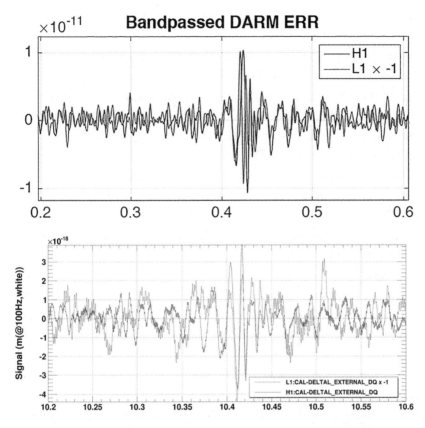

The first plots of GW150914's waveform (though not yet calibrated) were generated within hours of detection by Matthew Evans (*top*) and Stefan Ballmer (*bottom*). The strong signal enabled the waveform to immediately stand out. *Courtesy of Evans/Ballmer/Caltech/MIT/LIGO Laboratory.*

surprised that gravity waves from black holes, not neutron stars, were the first to be detected. Although black hole binaries are thought to be fewer in number, their larger masses produce a stronger gravity-wave signal than neutron star binaries when they merge. To Thorne, this fact upped the chances that black hole binaries would be seen first.) Yet the brevity of the signal—only one fifth of a second long—meant that the objects merging were heavier than neutron stars. The more mass involved, the greater the gravitational attraction, and hence the shorter duration of the system's final moments. If the signal was indeed real, the objects had to be black holes, but their masses were heftier than anyone anticipated. The binary coalescence software had missed the signal because it had been set to detect lower-mass neutron stars—not big, fat black holes. The coherent burst software detected the wave only because the signal was so strong. As a follow-up, the binary coalescence group went ahead and ran each observatory's recorded data through software that included black hole templates. This analysis, too, found the signal, and it agreed with the coherent waveburst group's initial finding: the signal indicated the collision of two black holes of surprisingly high mass. "I was wishing it would be black holes as they are totally Einstein objects—Newtonian gravity cannot explain them," said Weiss.

The plan was that during LIGO science runs, astronomers would be contacted immediately about any sizable event so that they could search for an electromagnetic counterpart to the gravitational wave signal—from radio waves to gamma rays. The gravity wave offers an early warning system to visually catch a stellar binary collision or supernova in the act. But that automatic alert had not yet been activated during this engineering run. Astronomers were notified two days later. Their reports turned up nothing of note either in the sky or within the astronomical databases—except in one case. The Fermi Gamma-Ray Burst Monitor in space did register a weak signal four-tenths of a second after the detection and emanating from the same general direction of the sky. This gamma-ray transient remains a mystery; other gamma-ray telescopes did not see a similar event. Some conclude that the signal may have been just background noise. In any case, black hole mergers were never expected to emit electromagnetic radiation unless an ample gas cloud was nearby (although theorists, upon hearing of this coincidence, immediately dreamed up alternative ways in which gamma rays might be released).

At first the September event was known as G184098, a tag the database automatically gave it. But within two days that label changed to its official name: GW150914, following an astronomical convention marking in sequence the year, month, and day the signal was detected—2015/09/14. (When gravity-wave observatories are recording more than one signal daily, each event will likely have an added letter at the end—A, B, C, and so on—in the order that they were detected over the day.) For more casual conversation, though, there was much traffic on LIGO's internal Wiki discussing possible nicknames. Because the candidate's intensity was so forceful, it was not surprising that the recommendations included "The Hulk," "The Big Enchilada," and "Grav Slap." One person suggested "Albert," since the signal arrived as physicists were about to celebrate the hundredth anniversary of Einstein's general theory of relativity. Another proposed "Dawn," to honor the dawn of gravitational wave astronomy. And a LIGO collaborator's young daughter submitted "Rainbow Unicorn," because she thought that the detection was so magical. In the end, none of these code names stuck. When not using its official label, everyone simply referred to the signal as "The Event."

As LIGO teams began the long process of double-checking the authenticity of their candidate, one particular possibility loomed large in their early efforts. Once the sun rose on the day of detection and Livingston researchers began arriving at work, the detector operator Parker told his colleagues about the brief call he had gotten during the night asking about the interferometer's status and a potential event. The audio was bad, but Parker thought that the caller's name was Eric. Who was Eric? Hearing about this strange phone call, LSU physicist Gabriela González wondered whether the caller had been a hacker checking on his or her plot. Did some "evil genius" break into the interferometer's computer system and plant a false signal digitally? Could he or she have actually *moved* the mirrors to physically mimic a wave? This was not an unusual thought. During the Big Dog event, such malicious schemes were explored as possibilities. And that was true for GW150914 as well. Such skulduggery was Rai Weiss's most dreaded nightmare. "The worst thing would have been to come up with a result that was fraudulent," he said. "Getting the hacker out of my head was very, very important to me."

Matthew Evans headed up the investigation—an ironic choice. When Mike Zucker had handled the same job several years earlier for Big Dog, Evans had been on the top of his list as a potential culprit. Evans knew the most about the two interferometers—their controls, all their undocumented features—to carry out such a hoax. So, there was no one better than Evans to sniff out a wrongdoer. He first got involved because he realized that the filtering algorithm he had devised to plot GW150914's first

Several members of MIT's LIGO team, wearing booties to keep the LIGO laboratory
on campus clean. From left to right: David Shoemaker, Rainer Weiss, Matthew
Evans, Erotokritos Katsavoundidis, Nergis Mavalvala, and Peter Fritschel.
Photo by BryceVickmark.com.

waveform could also be used to check the system for surreptitious manipulations. It allowed him to reconstruct the gravity wave's entire path—from
the analog machinery that registered the wave to its conversion into a
digital signal—and see if anything strange happened along the way. His
inspection came out clean. "It assured me that the digital system *was* be-
having properly—nothing malicious, nothing injected, nothing added,
nothing subtracted, nothing strange," said Evans.

But LIGO left no stone unturned. Researchers kept coming up with pet
theories on how such a dastardly deed could have been carried out. It was
Evans's job to check them out and (hopefully) shoot them down. At the
same time, all staff members and visitors at each observatory were polled
about their whereabouts and activities between 3:50 and 5:50 a.m. Central
Daylight Time on the day of detection. And once the interferometers were
unfrozen and the background run was completed toward the end of September, LIGO teams examined every nook and cranny of the instruments

to see if anything had been disturbed or altered, taking pictures along the way. Cybersleuths checked to see if tell-tale signs of a digital intrusion had been left behind in the computer records or if the timing of the computers had been altered in any way.

Effler and Schofield were prime suspects because they were the last to work around one of the interferometers right before the detection. Both were thoroughly grilled, as were other staffers. But more suspicion was focused on outsiders. Effler noted at the time that it would be difficult for a trespasser to enter an observatory unnoticed. "If somebody walked in, we would see it because of the seismometers and accelerators all over the place," she noted. "And if they had injected the signal into the suspension, it would have shown up in the suspension channel." Did anyone fiddle with the photodiode that directly records the gravity-wave signal? Effler was sure that didn't happen because she had crawled around in that area less than an hour before the detection and had seen no weird, extra equipment.

But maybe it was feasible to activate a hack offsite. The staff members most familiar with the interferometers were asked to do a variety of sanity checks, even to the level of opening up every electronics chassis to make sure that unusual hardware, such as a wireless communication device, had not been covertly inserted by a hacker. Said Caltech physicist Ryan de Rosa, "They asked me to look at the chips to make sure there wasn't something there to inject a signal," possibly activating the electronics via a cellphone from afar.

As it turned out, there was not one sign that a hack had occurred. And in the process LIGO scientists were reassured that such a fraud was virtually impossible. The more they thought about the technicalities of carrying out such a hack, the more implausible it became. "I was positive it wasn't an artificial injection," said Betzwieser, "because (a) I have intimate knowledge of how all this stuff works, (b) I know how hard it would be to fake it, and (c) it's too complicated. You'd mess up somewhere." Moreover, such a deception would require more than one evil genius. "You'd need an instrument expert at each site, plus somebody who's fluent enough in general relativity to generate the waveform," added Betzwieser. Even Advanced LIGO's blind injection team, whose full-time job was to generate such a false signal, had not yet done it successfully on the new equipment by the time the gravity wave arrived.

As hacker worries subsided, LIGO focused on the next crucial step in the examination of The Event. Was it truly astrophysical or just an instrumental hiccup? "We don't believe an online candidate is a true candidate until an offline analysis confirms it," said González. That meant a complete, unbiased assessment by a set of computer software searches. Some were based on general relativity, but others were not—a way of reassuring the team that they were not just seeing what they expected to see from relativity. Each analysis was carried out in a blind fashion with no human intervention: streams of data (that might or might not contain a signal) were analyzed, comparing them to the background data to estimate the probability that something in the data streams was of astrophysical origin. For days on end, servers in the United States and Europe ran constantly (in total some fifty million computer-processing hours) to reach their conclusions. The results were then put into a computerized "closed box" that could not be altered after the fact.

A suspenseful teleconference was held on October 5, 2015, exactly three weeks after the detection, to "open" the box. LIGO set a record on the number of collaborators who tuned in for this special edition of its weekly Monday telecon. Tito Dal Canton, a postdoc at the Albert Einstein Institute, said he "felt a shivering down my spine. . . . I couldn't wait to see the result. I wondered how the event would look through [my computer model's] glasses. Which part of the template bank would pick up the event? . . . The possibility that the event would not show up at all crossed my mind."

But Dal Canton's worries were for naught. Each search found a gravity-wave source at the very same point in the data, and each method displayed an identical waveform. More than that, the waveform was precisely what was expected for two black holes colliding, based on Einstein's equations of general relativity. The statistics indicated that the candidate was undeniably valid. The confidence level went beyond the science standard known as "five sigma," a measure that indicates 99.9999 percent certainty. An event like GW150914, it was later calculated, would happen by chance as a noise within the detectors less than once every two hundred thousand years. Champagne was soon pouring at universities and institutes around the globe (with the empty bottles taken home as souvenirs).

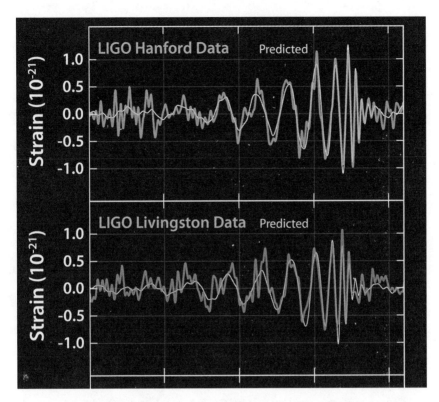

The thick lines indicate the gravitational waves captured over half a second at LIGO's Hanford detector (*top*) and the Livingston detector (below) on September 14, 2015. These plots, nearly identical, depict the periodic stretching and squeezing of space within each detector (the strain) as the waves passed through. Going from left to right in time, the waves at first show the two black holes orbiting each other faster and faster; they grow in intensity during the swift merger, and finally diminish during a ringdown as the new single hole settles down. The thinner lines show how the predicted waveform generated from general relativity closely matched the LIGO data. *Courtesy of Caltech/MIT/LIGO Laboratory.*

Was this the point that LIGO scientists *believed* that the signal was a real event? "It's not a matter of belief," answered de Rosa with a chuckle. "This is a physics experiment. It's a measurement."

So, opening the box marked the end of LIGO's preliminary evaluations but only the beginning of the detailed analyses of the measurement. Yet, one procedure was by then permanently crossed off LIGO's checklist: blind injections. Now that suitable candidates were registering, simulations were no longer needed to keep everyone on their toes. "We couldn't have asked for a more exciting reason to cease and desist!" joked ex-blind-injection team member Kissel.

Now came the hard work—assembling an airtight case for a discovery. A detection committee, headed by Stan Whitcomb (who had just officially re- tired), served as the devil's advocate. These committee members remained the skeptics as they reviewed all the evidence presented to them over the ensuing weeks. And that information was immense. For one, there were reports on all the environmental and physical conditions at each observa- tory around the time of detection, including cosmic-ray showers, wind speeds, earthquakes, airplanes flying overhead, magnetic storms, solar events, and broadcast radio interference. Fortunately, all of these potential spoilers had been quiet (or quiet enough), although there were some queasy stomachs for a brief moment when it was learned that an unusually large lightning strike had hit the Earth in the African nation of Burkina Faso dur- ing the very same instant as The Event. That's not unusual; there are tens of lightning strikes hitting somewhere on our planet each second. There were fifty-nine other strikes during GW150914, but the Burkina Faso bolt was one of the most powerful ever recorded. Further study, however, showed that all those coincident strikes were still at least a thousand times too weak to have any effect on LIGO. There were also two minor earthquakes off the West Coast within twenty minutes of GW150914, but fortunately the result- ing ground motion was ten times too small to have affected the detectors.

Meanwhile, data analysts continued their explorations, mining the chirp for all possible information on the nature of the event. And instrumental- ists kept reviewing the major equipment channels to catch any error that might have affected the detection's data quality. This extraordinary vetting was due to the signal's uniqueness—if it passed this tough scrutiny, GW150914 was going to be announced as the first gravity wave detected directly. The proof for such a watershed moment, like Caesar's wife, had to be above suspicion. (Once waves are detected on a regular basis and their profiles are more readily recognizable, the confirmation process will get more streamlined.)

By the end of October, gaining confidence, LIGO leaders had assembled a "paper coordinating team" to begin work on the scientific journal article. Peter Fritschel at MIT and Eric Chassande-Mottin at the Laboratoire AstroParticule et Cosmology (Astroparticle and Cosmology Laboratory) in Paris were chosen to chair this effort. Other members, added to the team to handle various sections, were specifically selected because of their skill in writing for a wider scientific audience. "Innocent graduate students are going to be forced to read this paper for the next several decades," Joe Giaime drolly noted. "So, we didn't want them cursing us under their breath because we introduced too much jargon." LIGO leaders decided that deeper details on the instrumentation and theory would be contained in a series of follow-up papers.

The first drafts went out to the entire LIGO and Virgo scientific collaborations, with their hundreds of members, for their input. This mass mailing generated a humongous surge of feedback, a situation that many compared to "herding the cats." Soon, the LIGO and Virgo groups devoted to various specialties within the enterprise (software, data analysis, calibration, computing, detector characterization, and such) were encouraged to have their members meet together, with each team submitting a unified response containing their corrections or suggestions. Some even held informal parties— going over each draft paragraph by paragraph and sentence by sentence.

But LIGO was still not on an easy glide toward its discovery announcement. There were the rumors LIGO had to confront.

It wasn't surprising that gossip about a possible discovery began circulating through internal physics grapevines fairly quickly. This chatter remained mostly discreet until September 25, when Arizona State astrophysicist Lawrence Krauss (who has no connection to LIGO) sent out a tweet to his more than two hundred thousand followers. "Rumor of a gravitational wave detection at LIGO detector. Amazing if true. Will post details if it survives," he wrote. Going public in this way, Krauss's message traveled around the world like an unfettered photon at the speed of light and was picked up by some in the science press. More details slowly leaked over the coming months. The steady drumbeat of this buzz put pressure on all the LIGO collaborators and affected their productivity; some began to refer to Krauss as "that a*stronomer." Sworn to secrecy, everyone associated with LIGO became quite skillful at feigning ignorance. "We really had to be good actors," said Livingston technician Gary Traylor. "I immediately began getting texts from friends but gave the standard response: 'We're collecting and analyzing data continuously.'"

Livingston operator William Parker was pained when his former Southern University physics professors asked him about the rumors. "No," he replied, with his fingers virtually crossed behind his back. "Nothing has happened. Wouldn't I know?"

But for some the subterfuge was thrilling. Cardiff University relativist Mark Hannam attended an astrophysics conference two months before the public announcement and noted that "it was a strange (and delicious) experience to have the secret knowledge of our discovery. There was a lot of talk about the potential of gravitational wave astronomy. . . . Do binary black holes exist, and do black holes exist with many tens of times the mass of the Sun? . . . [Collaboration] members in the audience already knew the answer!"

But at that stage, LIGO spokesperson Gabriela González wasn't as confident. The rumors placed the most stress on her. There was still a small chance that some overlooked bug might pop up, ruining their case. González would have been responsible for the public explanation. "It could have been a disaster—if we had to say why we *thought* we had something, but then we didn't," she noted. People might have rushed to judgment and declared LIGO a failure. At the time, she worried how such a backtrack might affect their future funding, especially given Joe Weber's rocky history and an infamous incident in 2014 when a group of astronomers prematurely announced the discovery of an indirect gravity-wave signature within the cosmic microwave background, only to retract the claim and have a part of their work discredited. That's why the LIGO collaboration was determined to say *nothing* until their scientific paper was peer-reviewed and accepted by a noted scientific journal. (Radio astronomers followed the same tight-lipped procedure when they found the first pulsar in 1967.)

From November into January, the detection paper went through eleven drafts, testing everyone's patience. "There were several things in the paper that were not my favorite choices," said González, such as how graphically to display the data. "That was okay with me, but others wouldn't let go." A joint

LIGO-Virgo collaboration teleconference was finally held on January 19, 2016, to decide whether to submit the latest draft, but because some reviewers would not yet sign off on the evidence, the vote was postponed. During a second meeting two days later, objections continued to be aired, such as what was the best method for calculating how often such a signal would pop up randomly due to noise. But then the majority rose up and voted to submit the article. Even before the experimental run began, LIGO collaborators had collectively chosen *Physical Review Letters* to reveal any discovery, because that journal had been the most supportive over the years in publishing LIGO's preliminary findings and technical accomplishments, unlike *Science* or *Nature*. "To avoid information slipping out from a casual conversation or a glance at a screen," noted *PRL* editor Robert Garisto, "we used the code name 'Big Paper'" to refer to the submission within the journal's offices.

Afterward, there was the nervous wait. "I still thought, in the back of my mind, 'Maybe we forgot something,'" said González. But the journal had agreed that the turnaround would be swift, only five days in the end, and the referees' responses were glowing. They had no doubts whatsoever that a gravity wave had been detected. "The reviews . . .," said Garisto, "conveyed the message that the paper would be an inspiration to physicists and astronomers alike." After LIGO collaborators generated a twelfth draft to address some minor referee comments and to insert last-minute additions, the paper was officially accepted on January 31. For Rai Weiss this approval was a huge relief; he no longer had to worry that, like some Pied Piper, he had led hundreds of people to the end of their careers. "The monkey that was on my back jumped to the sidewalk," he said. But Weiss still worried about all the work still to come, fixing LIGO's bugs and increasing its sensitivity: "That's why the monkey is still there, although he's not laughing as much."

González was also relieved, because she and a team of LIGO-Virgo leaders had already put LIGO's announcement plans into motion. All was scheduled to be revealed at a news conference on February 11 at the National Press Club in Washington, D.C., the same week when the annual meeting of the American Association for the Advancement of Science was being held in the capital.

* * *

LIGO Laboratory executive director David Reitze announces the discovery of GW150914 at a Washington news conference on February 11, 2016. *Caltech.*

David Reitze, executive director of the LIGO laboratory, confidently strode to the podium at the press club and after a brief pause declared with a smile, "Ladies and gentlemen, we have detected gravitational waves. We did it." The standing-room-only chamber resounded with loud cheers. Reitze continued after the long applause: "These gravitational waves were produced by two colliding black holes that came together and merged to form a single black hole about 1.3 billion years ago." That's just about the time that blue-green algae were the highest form of life on Earth. Having only two detectors, LIGO couldn't pinpoint the exact location on the celestial sky from which the signal originated—only to within an area covering some six hundred square degrees (the size of about three thousand full Moons packed together). That huge swath of sky was in the southern hemisphere roughly in the direction of the Large Magellanic Cloud. That's why Livingston, in the south, saw it first and more northern Hanford a fraction of a second later.

From left to right:
Gabriela González, Rai
Weiss, and Kip Thorne at
the discovery
announcement. *Caltech.*

At the news briefing, audience members heard an audio rendition of the
wave. The real signal had started at thirty cycles per second—a deep bass—
and swiftly rose in a fraction of a second toward a middle C, nearly 300
hertz. Sounding more like a fleeting thump, it was music nonetheless to
LIGO scientists' ears, the glissando they had been waiting decades to hear.
From that ephemeral yet sweeping tone, physicists were able to parse so
much information: the intensity of the wave indicated the distance to the
event; the speed at which the two black holes spiraled in provided data on
their masses and how they orbited each other. They could also discern how
much energy was lost in the collision.

One of the holes weighed thirty-six solar masses, the other twenty-nine
solar masses. In their last hundredths of a second of life, the pair were
whirling about each other at half the speed of light, until "two black holes
die, and one is born," noted Reitze. Yet the mammoth black hole created
from the mighty collision, after it swiftly settled down (what physicists call

the ringdown), was less massive than the sum of its two progenitors, ending up at sixty-two solar masses. That's because some of the mass, about three solar masses, was totally transformed into pure gravitational-wave energy. Following Einstein's famous equation $E = mc^2$, those three Suns' worth of mass turned into fifty times more energy than all the stars in the universe radiate in one second. This rapid conversion was the most powerful astronomical event ever observed until then. But by the time those waves traveled 1.3 billion light-years—their energy spreading in all directions into the universe—the tiny portion that passed through Earth had far less collective vigor. Space-time between each of LIGO's sets of test masses was stretched and squeezed by a distance only a thousand times smaller than the width of a proton particle. Yet LIGO was still able to sense that amazingly small motion. Put another way, it was like measuring the exact distance from the Sun to the star Alpha Centauri, four light-years away, with an uncertainty no bigger than the width of a human hair. "LIGO is the most precise measuring device ever built," stressed Reitze.

LIGO accomplished a number of firsts with its discovery: it was the first instrument (1) to directly detect a gravitational wave; (2) to clinch the existence of black holes; (3) to find proof that *binary* black holes exist (no one was sure before GW150914); and (4) to see that stellar black holes can be heavier than twenty-five solar masses, possibly formed in dense star clusters where smaller black holes (forged directly from dying stars) collide and build up mass. On top of all that, LIGO's detection allowed Einstein's general theory of relativity to pass yet another test with flying colors, demonstrating that the theory's predictions held up even in the extreme gravitational environment of a black hole. "I think there's a miracle of sorts here," said Weiss at the news conference. "[Einstein's] equations were written in 1915, and they've been tested in the weak fields—the field of the Earth, the field of the Sun, the field of the solar system—and these fields are tiny compared with what we have now. . . . And, nevertheless, the field equations seem to work, which is amazing." Those two black holes followed Einstein's instructions, stashed away within his theory, to near perfection. All in all, it was quite a scientific haul from an astrophysical signal that lasted only one fifth of a second. "This was bold . . . truly a scientific moonshot. We landed on the Moon," said Reitze in his closing remarks to the Washington audience.

An illustration of what two black holes about to collide would look like if you were able to get close to them. *SXS Collaboration, https://www.black-holes.org/.*

There were so many ways this gravitational wave could have been missed or rejected: if Weiss had postponed the scientific run to fix the radio interference problems, if Effler and Schofield were able to link with the GPS to do their car injection test, compromising the incoming data, or if Betzwieser and Kandhasamy had worked on the photon calibrator into the daylight hours. "If I had decided to be a workaholic, I would have prevented us from getting the detection. Sometimes it's good to go home," said Betzwieser. It's as if the gravity-wave gods became impatient after the LIGO team's decades of struggle and decided that there would be a detection, no matter what final checks on the instrumentation were left to be done.

Physical Review Letters released the journal article online the very moment the news conference started. "The demand for the paper was . . . so great that our site crashed under a load of ten thousand hits per minute," noted *PRL* editor Garisto. "After we added a slew of servers, our site came back up, and the paper was downloaded an unprecedented quarter of a million times on the first day." The paper's listed authors, more than one thousand, could have formed a mini-United Nations based on their country of origin. They worked in the United States and fourteen other nations. The work had been carried out at ninety universities and research institutes situated around the world. The reason for listing so many authors was unassailable: the discovery was not made by the people present at the observatories that auspicious September day; the discovery arrived only after a battalion of researchers spent decades designing the infrastructure, developing innovative instrumentation, achieving breakthroughs in numerical relativity, and writing miles of computer code to carry out the data analysis. "It was the most important scientific paper in the careers of a thousand people," noted Mike Zucker.

Weiss once posted on his MIT webpage that he "started in physics with the precept that only very important and finished work should be published. . . . As a consequence, didn't publish much—and got hell for it." When the discovery was announced, MIT president L. Rafael Reif said that he appreciated Rai's high standards, "and if you *had* to publish something, this one was an excellent choice."

According to many of the LIGO researchers, they were lucky that the first bona fide gravity wave happened to be so strong. A weaker signal would have made a shakier case, likely leading to doubts within the physics community and the need for further observations before anyone celebrated. Before this, "if you asked most of us what our first paper would have looked like," commented Giaime, "it would have been a paper that carefully analyzed why this barely visible signal we saw wasn't fake." Instead, the various LIGO teams were able to write a multitude of papers dipping deeply into the rich science of the event.

While the discovery announcement was under way in Washington, LIGO collaborators viewed the proceedings via teleconference during their own celebrations at universities and institutes around the globe. Upon hearing David Reitze's opening words, said Gary Traylor at Livingston, "Cold chills immediately overcame me. It was a proud moment. I wanted to stick my chest out because I was a part of this." He has been with LIGO since 1999, just when they were finishing up construction.

For an entire press cycle, the announcement dominated the news in print and on cable, radio, and television. The charmed name Einstein brandished its enchantment once again. "We actually rose above the presidential election," said Betzwieser. "That's saying something—science beat out politics for a day!" And the public responded in kind. When the Livingston observatory held its monthly "Science Saturday" nine days after the announcement, around 1,300 people showed up for the educational fair, far more than usual. There was mayhem and heavy traffic, with people parking everywhere. Fashion designers even took notice of the discovery. In California, Holly Renee, who specializes in science-themed clothes at her company, Shenova, quickly manufactured a digital fabric that featured crisscrossed images of the gravity wave, which was used to make a stylish sheath dress.

So revered was this discovery within physics circles that major honors arrived quickly, possibly setting a record. Within four months of the public announcement of GW150914, the troika—Rai Weiss, Kip Thorne, and Ron Drever—received four prestigious awards: the Shaw Prize in Astronomy, the Gruber Foundation Cosmology Prize, the Kavli Prize in Astrophysics, and the Breakthrough Prize in Fundamental Physics. The three scientists were cited for their key contributions in founding LIGO, which led to the first observation of gravitational waves.

For another LIGO scientist, there was a more minor award (of sorts). A month after the discovery announcement, Lisa Barsotti calculated who among the LIGO collaboration had devoted the greatest fraction of their existence—up until September 14, 2015—on discovering a gravity wave. Surprisingly, it wasn't Rai Weiss, who had not gotten into the game until his forties. The winner was Mike Zucker, who started at the age of nineteen in 1980 as a freshman at the University of Rochester. There he worked with David Douglass on the physics department's bar detector. Afterward, Zucker never left the field, even heading up the Livingston observatory at one point. "I had invested something like 65 percent of my time on Earth to gravity waves until the detection," said Zucker. "Unlike me, Rai had a life before that."

But GW150914 was hardly LIGO's only concern during this time. As the paper for first detection was being prepared and staff members were conducting their verifications, LIGO's science run was still continuing. A number of gravity-wave triggers occurred throughout this period; most were equipment bugs or indistinct signals that failed to meet the mandatory standards for selection.

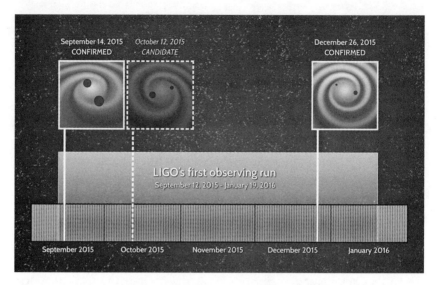

Timeline: the three detections during LIGO's first four-month science run. *Courtesy of Caltech/MIT/LIGO Laboratory.*

Identical control rooms at Livingston and Hanford have desks filled with computers and walls covered with monitors, allowing the interferometers' operators to keep watch twenty-four hours a day, seven days a week. *Courtesy of Kimberly Teske Fetrow.*

LIGO had been planning to end O1 just before the December holidays, but with the interferometers working so smoothly and ongoing delays in finishing the *Physical Review Letters* paper, LIGO leaders decided to continue the run until January 12. It was a fortunate decision, because on Christmas Day, Santa delivered a very nice present to each observatory—their second viable signal, another pair of black holes colliding. Livingston detected the gravity wave first (at exactly 9:38 p.m. and 53 seconds local time), with Hanford registering the event one thousandth of a second later. But as this event occurred in universal time (LIGO's official timeclock) at 3:38 a.m. the next day, it became known as the "Boxing Day Event," named for the day after Christmas in Great Britain when employees or tradespeople traditionally receive gift boxes from their employers or customers. So, officially, this gravity-wave event became GW151226.

This signal was not as strong as the September event, because it involved the merger of smaller-mass black holes—best estimate, fourteen and eight times the mass of the Sun. The single black hole that resulted weighed twenty-one solar masses. That meant that roughly one solar mass was wholly converted into gravitational waves, which proceeded to disperse into the universe and arrive at Earth 1.4 billion years later. Although the holiday signal was weak, there was an upside to this collision. "Because of their lighter masses compared to the first detection, they spent more time—about one entire second—in the sensitive band of the detectors," noted González. This allowed LIGO scientists to scrutinize twenty-seven orbits before the black holes' tremendous clash, more than five times the orbits observed during GW150914's meager fifth-of-a-second window.

The immediate reaction within LIGO to the Boxing Day Event was interesting compared to the response to GW150914. There was joy indeed, but the remarks within LIGO's electronic logbooks were more subdued. After their painstaking experience with GW150915, everyone was more confident in their follow-ups to this event, more matter-of-fact. Even the press got more blasé. Where GW150914 was front-page news, articles on GW151226 (made public in June 2016) got shuffled off to the inside pages.

Advanced LIGO received one more tempting trigger during its first science run, tucked in between the run's two verified signals. This event occurred on Columbus Day, October 12, 2015. Because it also occurred at the start of a workweek, like GW150914, everyone called it the "Second Monday Event." This signal also appeared to be two black holes colliding, but its very weak strength meant that scientists could not confirm it as a gravity wave with adequate assurance. So, they named it LVT151012 (for LIGO-Virgo Transient). A Twitter page dedicated to this wannabe sends out a plea to help it "climb out of the background," which it might someday as the field matures and gravity-wave detectors have built up a good databank of signals for comparison.

But at the time LVT151012 served as a great psychological boost. Throughout September LIGO was dealing with only one, highly unusual candidate. "A whopper without the fries," joked Zucker. LIGO scientists expected signals to come in a distribution—with every big signal followed by a bunch of little ones, but for weeks the interferometers detected nothing of significance. So, it was a relief when the Second Monday Event arrived, calming everyone down that the first event was not a one-of-a-kind glitch after all. The first science run for Advanced LIGO came to follow a script resembling a cosmic version of "Goldilocks and the Three Bears"—first the big event (GW150914), then the small event (LVT151012), and finally one in the middle (GW151226).

Variations on a Theme

PISA, A VITAL PORT IN ROMAN TIMES, rises thirteen feet above sea level on the banks of the Arno River. The city was a seagoing power in the twelfth century, when it became a republic after participating in the First Crusade. At that time its influence extended over the entire coast of Tuscany, sparking economic prosperity and artistic splendor. It was during this period of affluence, in 1174, that Pisa's campanile—what became the famous Leaning Tower—went under construction: six tiers of arches on a base, each precariously perched one upon the other and topped with an airy belfry. Overall, the white marble tower resembles a too-tall wedding cake about to topple, as if its builders were on a drunken engineering binge. The lean started as the third level was being built. The ground began to sink owing to a water-bearing layer underground. A cultural icon today, the structure is continually surrounded by tourists, their cameras and selfie-stick-held phones clicking away.

It is here, the legend goes, that gravity first came to be understood in a scientific manner. Although the story is likely apocryphal, it is said that Galileo dropped balls of various weights from the top of the campanile to prove his new view of gravity. Until then Aristotle's word was the standard law of physics when it came to falling objects. The Greek sage had declared that the heavier a mass is, the faster it falls to the ground. But from tests he conducted, Galileo concluded that this long-accepted fact wasn't true. He figured out that the duration of a fall is independent of the mass.

Neglecting air resistance, a tiny marble will fall just as fast as a heavy bowling ball. In his *Dialogues Concerning Two New Sciences,* Galileo had one of his characters, Sagredo, describe the test: "I . . . who have made the test can assure you that a cannon ball weighing one or two hundred pounds, or even more, will not reach the ground by as much as a [hand] span ahead of a musket ball weighing only half a pound." Galileo begat Newton, who begat Einstein. Our modern understanding of gravity began with that one elegant thought.

The leaning tower had been closed off in the 1990s. A series of weights were attached to its base on the north side to counteract the ever-increasing lean. As officials in Pisa worked to save the legendary gravitational test site, others nearby were pushing gravitational research into the future. A French-Italian collaboration built a LIGO-like detector, known as Virgo, on the vast alluvial plain just outside Pisa. Although LIGO is quite capable of making a gravity-wave discovery on its own, since its two separated sites offer the opportunity to reject local disturbances, the biggest scientific payoff arises when LIGO is operating as an element in an international network of gravity-wave detectors. Virgo is part of this network, as are other detectors. Near Hannover, Germany, a British-German team operates GEO600, an interferometer that is six hundred meters, or almost two thousand feet, long. (Virgo didn't detect GW150914 because it was then offline, in the midst of having its second-generation instrumentation installed. And GEO600 did not have enough sensitivity.) Japan, meanwhile, is about to commission an underground detector called KAGRA (for Kamioka Gravitational-Wave Detector) that will use supercooled mirrors held within arms that are three kilometers long, or almost two miles. And India has plans for a LIGO-like detector that will be built in collaboration with the United States.

Together, these facilities will form a worldwide system, akin to the global network of radio telescopes that allows radio astronomers to coordinate their observations. Experts imagine there will eventually be a range of detectors of various designs, each providing a unique contribution toward helping understand a gravity-wave signal. Detection depends on them all "listening" and comparing notes. The direction to an abrupt bursting source, such as a binary collision or supernova explosion, cannot be adequately

Locations of the gravitational-wave detectors now in operation (LIGO Hanford and Livingston, Virgo, GEO600), undergoing commissioning (KAGRA), or planned (LIGO India). *Basemap data © OpenStreetMap contributors (CC BY-SA).*

achieved with fewer than three or four observatories, the way a surveyor needs several points to peg a position. Each site acts like a surveyor's stake in triangulating the location. With LIGO and Virgo working together, it will be possible to pinpoint a source to within several square degrees on the sky. When more detectors are added to this network, the resolution will improve even more. (A single observatory will still have the opportunity to determine the general direction of a continuous gravity-wave source, such as a pulsing neutron star, by noting the Doppler shift in the signal's frequency as the Earth moves in its orbit about the Sun.)

The Virgo project began as a joint effort that involved some hundred physicists and engineers from Italy and France. But more than a hundred additional European collaborators from the Netherlands, Poland, and Hungary have since joined, a consortium known as the European Gravitational Observatory, with Virgo as its detector. Construction of the interferometer began in May 1996. Its arms are a bit shorter than LIGO's, three kilometers instead of four. Set on flat farmland of clay and sand, these arms extend from south to north and from east to west. The surrounding farmers raise sugar beets, corn, and sunflowers for oil. To the north are the mountains from which Michelangelo obtained the marble for his sculptures.

Before laying down the pipes over those many miles, Virgo engineers had to check for lost mines from World War II. One of the war's most important frontal assaults passed right through this site along the Arno. From afar the central complex looks like a small industrial park. A lengthy ditch, bordered by earthen dikes, runs by the buildings. This channel serves as an

The Virgo interferometer located south of Pisa, Italy. *Virgo Collaboration.*

emergency runoff should the nearby river overflow. In fact, Virgo was plagued by frogs during its construction, owing to water in its basement.

Adalberto Giazotto codirected the Virgo project in Italy when it was under construction. His counterpart in France was Alain Brillet. Giazotto, sometimes called the father of Virgo and now retired, still consults on the endeavor from his offices located at the Istituto Nazionale di Fisica Nucleare (National Institute for Nuclear Physics), or INFN, south of Pisa, in a quiet suburb called San Piero a Grado that is surrounded by a national forest and just a short drive from the Virgo detector. "Here is where Virgo was born," he said with pride as he arrived at the laboratory. The name refers to the Virgo cluster of galaxies; the project's initial aim was to create an instrument that could detect supernova explosions as far out as that noted collection of galaxies, some fifty million light-years from Earth.

Giazotto is tall and trim, with thinning silver-colored hair and the bearing of an aristocrat, but his office is decidedly utilitarian, with its one desk, one cabinet, and one set of file drawers. Right outside this spartan office is a walkway that looks down on a laboratory where the Virgo detector equipment was readied in the 1990s. Like many in this field, Giazotto came over from particle physics. An experimentalist, he worked with synchrotrons to study the weak nuclear force and the structure of nuclear particles. But he saw no great difficulty in jumping to the study of gravity. To him it was still a particle physics problem. "To discover the graviton," he declared was his goal, the theoretical particle that transmits the force of gravity the way a photon transmits the electromagnetic force. "The only way to see them is to build an observatory. And then there's the enormous bonus of better understanding the universe."

He started thinking about gravitational-wave detection in the mid-1970s while he was at CERN working on particle physics experiments. By the next decade he was actively campaigning to get Italy into the business of laser interferometers. But he had a decided point of view. From the outset he wanted to build an instrument that could detect much lower frequencies than the other systems in the works, and that meant focusing on the problem of seismic isolation, the largest impediment to detecting low frequencies. He presented the results of his first tests in the mid-1980s at an annual conference on gravity then meeting at the University of Rome. There he met

Brillet, a French pioneer in laser interferometry, who was also interested in building a large system. Eventually, they teamed up and arranged for support from the physics communities in both countries. Giazotto firmly believed that detecting lower frequencies was vital for studying certain sources. Take coalescing binary neutron stars, for instance. "From 100 hertz and upward, you have just three seconds of observation time before the stars coalesce," he noted. "But if you start at 10 hertz, you can get a thousand seconds of observing. That's why I [wanted] the low frequencies." His goal was for Virgo to get down to 4 hertz. In terms of wavelength, that means gravity waves that span almost forty-seven thousand miles from peak to peak (one-fifth the distance from Earth to the Moon). The first-generation Virgo achieved that frequency target before it was shut down for an upgrade, similar to the one that LIGO's instrumentation underwent—larger mirrors, more laser power, and signal recycling.

To get to 4 hertz, environmental motions of the suspension system had to be reduced by many orders of magnitude. The trick was to stop the outside vibrations from flowing down the wires that hold the masses and thereby jiggle them. To do this, the Virgo researchers devised a unique seismic isolation system called the super attenuator. It has six circular rings, stacked one on top of the other to form a structure three stories tall. In certain ways it resembles the multitiered Leaning Tower, without the lean. Each of these rings is a mechanical filter, consisting of six triangular metal blades under enormous tension, enough to block the noise flowing down the wires to the test mass suspended at the bottom. The scheme worked, even on the prototype first erected in the INFN laboratory. "We shook the top with a motor to produce a displacement of about one millimeter at 10 hertz," says Giazotto. "At the bottom we couldn't detect any change, at least to a level of 10^{-10}." Ordinary seismic motions, the kind the instrument would face daily, are actually less than that magnitude.

Giazotto had always dreamed of erecting a second Virgo nearby. With two detectors close to each other, the detected waves would be in phase, rising and falling in concert. By combining them, one would get a decided boost in signal. This pairing never happened, but in some ways Virgo already has a mate. A second laser interferometer has been operating in Europe for some time, the product of one of the world's earliest programs in laser interferometry that is centered in Germany. This effort in gravity-wave detection originated under the guidance of Heinz Billing at the Max-Planck-Institut für Physik und Astrophysik (Max Planck Institute for Physics and Astrophysics) in Munich. (Billing, in his later years, was once asked what he would yet like to witness in his lifetime. He answered, "To see the detection of gravitational waves." He reached that goal on September 14, 2015, at the age of 101. He was a year old when Einstein introduced his general theory of relativity.) Billing and his colleagues came to the idea almost by accident. At the time their institute included a special physics division devoted to building computers for scientific calculations, a vital need in the days when commercial computers did not yet offer the power to handle complex scientific equations. But as the computer industry caught up to these needs, the institute's computer designers found that they were losing their mission. They gained a new vocation when Joe Weber announced that he had detected gravity waves. "The astrophysicists got very excited by this," recalled Roland Schilling, a former member of the computer-building team. "They said it was so exciting that it would revolutionize all of astrophysics if the claim were true." Theorists on staff at the institute wanted to repeat the experiment, but only the computer development group had the necessary laboratory expertise. Almost overnight its members became detector builders for gravity-wave astronomy.

The Munich group soon had a room-temperature bar up and running at its facility. The group coordinated its operation with another bar independently built in Frascati, Italy. Within a few years, though, it became evident that Weber's results could not be confirmed. But a spark had ignited. By then there were two alternatives to improve performance: either construct a supercooled bar or switch to laser interferometry. The Munich researchers, wanting to stay in the business but having neither sufficient expertise nor the necessary infrastructure in cryogenics, opted to go into laser interferometry. They were inspired by the work of both Rai Weiss and Bob Forward. They ordered their first laser in 1974 and were highly optimistic. "I remember very well that we originally had a very short timescale in mind. We thought it would take five or ten years," said Schilling. It all looked so promising on paper.

But just like Ron Drever in Scotland, the Germans quickly discovered the many pitfalls of the new approach. Their first mirrors were rather bad; their

poor surface quality scattered the light, which greatly reduced the sensitivity. They spent years learning how to deal with laser beam jitters as well as
figuring out the best way to mount the test masses. At first the mirrors were
just clamped onto aluminum blocks, but such a linkage was a major source
of vibration. Their first interferometer was a mere thirty centimeters (about
a foot) long. Afterward, they went to arms three meters (about ten feet) in
length. Despite the technological challenges, they remained encouraged,
especially when Joseph Taylor came to a meeting in Munich in 1978 and
announced his first gravity-wave results from the binary pulsar. Here, at
least indirectly, was proof that their elusive goal was not imaginary.

Throughout the 1970s and into the 1980s the German laser interferometer team held all the records for sensitivity for such an instrument. The
Munich group got a fantastic head start by using Weiss's initial interferometer design and then reducing everything to its bare bones. Along the way
they made many valuable contributions to the technique, such as learning
to suspend the masses on wire slings to reduce seismic interference and
thus let the mirrors themselves serve as the test masses. These innovations
are now part of the standard instrumentation in every interferometric
gravity-wave observatory either planned or built, but at the time the group
was immersed in the Model T era of the field, working out the basics now
taken for granted.

Since those early days, a time when researchers knew every worker in the
field by name, the community has grown to encompass hundreds of engineers, technicians, astronomers, and physicists. Technology was greatly
advanced, improvements now applicable to other arenas of physics. "Even
if we detect only the sources we expect," said Schilling, "the scientific gain
would be more than worth the effort." But there was another major reason
Schilling stayed in the field until his death in 2015. "If you compare what
has been experienced in optical, radio, X-ray, and gamma-ray astronomy,"
he said in 1998, "there were always sources that you did not think of before.
There were always surprises. Why shouldn't that also hold true for the
gravitational-wave business?" he said. It is a sentiment that has become the
standard refrain for the field, its raison d'être.

A turning point for laser interferometry came in 1982. From his work
with interferometer control, Schilling came up with the idea for power

recycling. At first he thought it wouldn't offer much benefit when incorporated into their design because it required tremendously good mirrors. But Ron Drever was aware that supermirrors were emerging from military technology and independently discovered the same principle: letting the light stay trapped, so that it continues to bounce between the mirrors. This boost makes it appear that the system is using a more powerful laser, which assists in decreasing a major noise in the system. Until this breakthrough, laser interferometry had been the poor cousin to bars in the gravity-wave game. Low laser power, scattered-light problems, innumerable vibrations, and poor mirrors meant that the technique remained a dark horse. But by the mid-1980s its status had vastly changed. The introduction of power recycling, along with supermirrors and stable lasers, was a critical development. "From the beginning the advantage of the laser interferometers was that they were broadband detectors," Schilling pointed out. In other words, they can register a wide band of frequencies, whereas a bar is confined to a narrow frequency range.

In 1983, now as part of the Max-Planck-Institut für Quantenoptik (Max Planck Institute for Quantum Optics) in Garching, a suburb of Munich, the team began to operate a laser interferometer with arms thirty meters long, or almost a hundred feet. After focusing on bettering its sensitivity, the team switched its emphasis to more technical issues, using the instrument as a test bed for new technologies. "Problems that had to be solved if you wanted to operate a big instrument," said Schilling. As in the United States, a much larger facility was very much on their minds. At first they thought they would naturally progress up the orders of magnitude. They had already built prototypes with thirty-centimeter, three-meter, and thirty-meter arms. It seemed reasonable to try three hundred meters, or a little less than a thousand feet. But, politically, they knew that they had to jump to three kilometers, or almost two miles, for that was the size where they'd have a chance to actually detect something, "even though it was against our common sense," said Schilling.

Rai Weiss's attitude was beginning to take root around the world: let's stop playing with the technology and get some results with a giant detector. While the Germans were thinking of that approach, researchers in Glasgow were pitching the idea for a similar facility to be built in Great Britain. With both countries facing budget crunches in the 1980s, the two groups joined forces in 1989. They named the project GEO, which loosely stood for Gravitational European Observatory. "I said we should call it EGO, for European Gravitational Observatory," noted Schilling with a smile (a suggestion politely ignored until it was later adopted as the observatory name for the Virgo detector).

In the summer of 1989, Herbert Walther, then managing director of the Max Planck Institute for Quantum Optics, met Karsten Danzmann at a laser spectroscopy conference and convinced him to take over the gravitational-wave detection effort in Garching. Danzmann had never been involved in gravity-wave physics, although he did have a habit of changing fields every few years. Trained in gas discharge physics, he worked for ten years on heavy ion collisions, laser spectroscopy, and the properties of positronium. He also had experience building ultraviolet lasers. But he readily accepted the invitation to try this new line of research. Gravity-wave research takes a person who is "stubborn, will take a risk, and is hopelessly optimistic," he said in his perfectly accented English, acquired during a stint on the faculty of Stanford University. There is a very subtle boundary, he added, between genius and being over the edge. "Let's just say that the percentage of people who are on the other side of that edge is a lot higher in this field than in any other," he said with a laugh. Danzmann felt right at home.

When he arrived at Garching, plans for GEO were well under way. But within a year the project fell apart. The financial difficulties faced by Germany in the aftermath of German reunification made funds for the endeavor disappear. By 1991 support was cut off. Great Britain also froze funding because of budgetary problems. All hope appeared to be lost, until Herbert Welling, a professor of physics at the University of Hannover in Germany, convinced his university to broaden the scope of its research efforts by setting up a chair in gravitational-wave physics, just as Caltech had done. Danzmann took the post in 1993, setting up an outpost of the Max Planck Institute for Quantum Optics in Hannover and transferring many on the Garching research team to the northern town.

Once in Hannover, Danzmann began talking with James Hough, who had taken over Drever's position as head of the Glasgow gravity-wave detection team, to resurrect GEO but on a much smaller scale. Instead of three-kilometer-long arms, this interferometer would have six-hundred-meter-long arms, a fifth the size of their original plan. "We refused to die," said Danzmann. They were able to get ten million marks, less than a tenth of the proposed cost of the original GEO project. "We were sure we could do it for very little money, if we were willing to take risks and do everything in a very unconventional way," he noted. "We just needed a bit of ingenuity."

Because of its six-hundred-meter-long arms (about two-fifths of a mile), the instrument was renamed GEO600. Danzmann claimed they had an advantage in being small. It has enabled them to be highly flexible. Equipment can be changed, almost as fast as a new idea is developed, unlike the larger systems, where designs must be chosen and locked into years in advance. This has allowed them to try out interesting new schemes, some of which were later adopted by LIGO and Virgo in their second-generation upgrades. They include stable, high-power laser systems and the multi-stage mirror suspensions using silica wires.

Despite its shorter arms, GEO600 always had the potential to attain a respectable sensitivity. Its initial goal in the 1990s was to detect strains along the order of 10^{-21}. It's now capable of doing ten times better than that at high frequencies. (When both LIGO and Virgo were down for several years to install their latest upgrades, GEO600 did primarily operate as a detector, so that the community wouldn't miss any big gravity wave–generating event if it had occurred nearby.) But more important now is using GEO600 as a testbed for advanced interferometer design. "If you're poor, you have to be smart to survive," said Danzmann. This means that they have a smaller chance to detect signals than their bigger cousins. Yet now that signals have been detected with a larger system, a host of smaller observatories, like the low-cost GEO600, could possibly be built around the world to enhance the gravity-wave network.

GEO600 is located south of Hannover, about thirty minutes by car. It's an agricultural test site, land owned by the state government and operated by the University of Hannover for agricultural research. The complex was erected right in the middle of working fields of wheat, barley, apples, pears, raspberries, strawberries, and plums. The arms were conveniently built along existing farm roads. Construction began in the fall of 1995, with a toast of single-malt whiskey, a nod to the Scottish connection. A few drops were sprinkled on the site. "German beer was applied internally later," said Danzmann.

The size of the instrument was dictated by the width of the land available. It might have been called GEO573. After 573 meters one arm hit the boundary of the allotted land. To make it an even 600, the collaboration for a while leased the last few meters from the farmer next door at a cost of twenty-seven pfennigs per square meter per year. That was about 270 marks annually, until the university acquired the plot through a land swap. Researchers have no plans to make GEO600 longer, turning it into a LIGO. Two hundred meters farther down the arm is the Leine River. "It's an experiment," stressed Danzmann. "It was not meant to last half a century."

If LIGO were described as an extravagant Broadway show, GEO600 might be called a high school play. To keep costs down, the German-Scottish collaboration depended heavily on technicians on staff at Hannover as well as student labor. "The entire central building is about as big as LIGO's electronics workshop," said Danzmann with a rueful smile. "Contractors did the heavy work, like pouring the concrete and putting on the roof, the bare bones building. We did everything else." Students designed the air-conditioning system for the clean room. It cost twenty thousand marks. A commercial system would have cost a million. They also saved money by going with risky designs, such as their vacuum beam tubes, which extend to the north and east.

"It's an unproven design that has never been built before. It was rejected by LIGO as being too risky, even though it costs about a tenth of others. We're using a vacuum tube design normally used for air ducts and air-conditioning systems. It's a thin-walled tube, which is stiffened by giving it a corrugation, all the way along. It's like a long bellows. Otherwise, it would collapse. It only has a wall thickness of 0.8 millimeters. It's a sixty-centimeter-wide tube. That way you use very little material, so the material cost alone is low, plus the tube is very lightweight, so handling is very easy. We can use students to carry it around. And the tube supports are a lot less demanding because the weight of the tube is like a wet curtain roughly. Heating of the tube is also easy because the walls are thin. Just a couple hundred amperes of electricity makes it hot," said Danzmann.

GEO600 also greatly enhanced its sensitivity with a technique called signal recycling, which was so successful that it, too, was adopted by LIGO and

Virgo. Glasgow physicist Brian Meers was the first to fully work out this idea. You might think of it as an interferometer acting like a bar, being tuned to a particular frequency. Consider a radio. You tune into a certain station with your dial, which locks your radio onto a carrier wave at a set frequency. The radio then ignores this carrier wave and looks at its sidebands, the frequencies right next door, where the music and talk reside. A gravity-wave telescope in some ways works similarly. The laser light is a very exact frequency. But if a gravity wave passes by, it moves the mirrors and affects the frequency of the laser light. The signal, in a way, is the music placed on either side of the laser light frequency. So the light, as it's being circulated within the interferometer, has these sidebands where the gravity-wave signal information resides. In signal recycling these sidebands are stripped off the laser carrier wave and sent back into the interferometer, which builds up and amplifies the signal. "That's the whole purpose of this instrument," said Danzmann, "to push the limits of what you can do experimentally. Of course, if all goes well, it may actually see a signal, although that is not its main purpose."

Other large laser interferometers are either under construction or in the planning stages. Japan initially had an interferometer known as TAMA-300. This detector was located in Mitaka, twelve miles from Tokyo, at the country's National Astronomical Observatory. Unlike the other detectors, the three-hundred-meter-long arms of TAMA were housed in long, concrete tunnels set completely underground, which diminished the environmental noises that can plague detectors on the surface. It was used as a test-bed to prepare for Japan's current venture, a three-kilometer-arm detector known as KAGRA, presently being built deep within the Kamioka mine. This interferometer will also use mirrors cooled to near absolute zero to reduce thermal noise to a whisper.

And spurred on by the first gravitational-wave discovery in 2015, India accelerated its plans to collaborate with the United States in building its LIGO observatory. The Indian cabinet granted approval just three days after the black hole collision GW150914 was publicly announced. As mentioned earlier, India will be installing equipment that was once part of a two-interferometer system at LIGO's Hanford observatory. LIGO scientists have already made dozens of trips to India to advise the key science institutes that will be responsible for constructing and operating this detector.

Yet even with these additional facilities, gaps remain in the layout of the observatories. All the laser interferometry endeavors are in the northern hemisphere, which somewhat restricts the geometry needed to pinpoint the location of a source on the celestial sky. With that in mind, researchers hope that the recent detection of gravity waves will spur ongoing plans to build detectors in the southern hemisphere as well, achieving more accuracy when it comes to cosmic triangulation. "A southern hemisphere detector is a major component of a global array," said John Sandeman of the Australian National University in Canberra. "One interferometer is not really a telescope. An observatory really requires at least a group of four interferometers."

And it doesn't end there; gravitational-wave specialists are already proposing the construction of third-generation detectors that could be up to a hundred times more sensitive than current instruments. A collaboration of several European institutions is studying a concept called the Einstein Telescope that proposes a set of detectors with longer arms (possibly ten kilometers, or about six miles), a location underground to cut down on seismic noises, more powerful lasers, supercooled optics as used in KAGRA, and nested interferometers (one detecting high-frequency waves, the other focused on low-frequency signals). Researchers in the United States are beginning to think about successors to LIGO as well, possibly with forty-kilometer-long arms (about twenty-five miles). If LIGO is equated with Galileo's original telescope, such advanced systems would be a leap toward modern-day scopes, opening up regions of the universe still hidden to current instrumentation.

The Music of the Spheres

THE PYTHAGOREANS WERE AN ANCIENT BROTHERHOOD founded by the Greek philosopher Pythagoras, who is best remembered for his famous theorem concerning right triangles: the square of the hypotenuse of a right triangle is equal to the sum of the squares of the other adjacent sides. In the fifth and sixth centuries B.C.E. Pythagoreans were devoted to such examples of mathematical beauty and extended this fervor into their contemplation of the cosmos. They intensely believed that celestial spheres ascended from Earth to heaven like the rungs of a ladder, which majestically carried the planets around and gave forth harmonic tones that created a wondrous music of the spheres. One version of this system had each planet intone a note higher than the one before it, starting with the Moon and working outward to the fixed stars. The seventeenth-century astronomer Johannes Kepler was a devotee of this idea and even wrote down some celestial tunes, audible to God alone, that he associated with various planets in their orbital journeys. Little did he realize that his musical vision would find substance in an entirely different astronomical arena.

For most of astronomy's history, the universe was studied with one means and one means only: collection of its electromagnetic radiation. For the astronomers who came before Kepler, it was with their eyes alone. Later, lenses and mirrors focused and magnified the visible light. By the middle of the twentieth century, astronomy had expanded its arsenal of instruments to collect photons from other regions of the electromagnetic spectrum—radio, infrared, ultraviolet, X-rays, and gamma rays. At the same time, particles such as cosmic rays and neutrinos began to be gathered from space. With each new technique, noted Penn State theorist Sam Finn, astronomers found something they didn't expect. Radio astronomers found pulsars, quasars, and massive molecular clouds dotting the celestial landscape. X-ray astronomers were surprised by the power of X-ray binaries, which strongly hinted at the existence of black holes. The lesson to be learned, said Finn wryly, is that "astronomers have no imagination."

Given that understanding, who knows what gravity waves might eventually uncover, for they do offer a far more radical method of gathering information about the universe. That fact is quite apparent when comparing electromagnetic radiation to gravity waves. Electromagnetic waves are emitted by individual atoms and elementary particles. Gravity waves, in contrast, are generated by the bulk motions of matter. The frequency of the wave is directly related to the frequency of the massive movement generating the gravity wave. At the moment, for example, the two neutron stars in the Hulse-Taylor binary are emitting gravity waves at a frequency of 0.0001 hertz as they circle each other every eight hours. As they get closer together over the millennia, though, and their orbital velocities speed up, the gravity-wave frequency will increase. LIGO will detect the waves from such neutron star systems when they match (coincidentally) the lower portion of the audio-frequency range (20 to 20,000 hertz).

The strongest gravity waves of all are created when matter approaches the speed of light, which occurs in supernovas or when black holes collide. This offers the opportunity to explore the most violent phenomena in the cosmos. It allows astronomers, for instance, to peer into the very heart of an exploding star, something that telescopes can't do. This is because gravity waves pass through matter as if it were not there, unlike most electromagnetic radiation, which can be absorbed or scattered by matter as the radiation proceeds on its journey. Electromagnetic waves travel through space-time, whereas gravity waves are actual jiggles of space-time itself. With gravity-wave detectors, we are learning about the cosmos from its space-time vibrations. As pointed out earlier, when these waves are played as an audio signal, they add sound to the many pictures astronomers have been gathering over the decades. Gravity-wave astronomers are listening to the modern-day version of Pythagoras's music of the spheres.

As Joe Weber and other experimentalists made their first forays into gravity-wave detection, theorists were active as well. As the number of detectors grew, specialists in general relativity launched their own effort to figure out what "tunes" the detectors might be receiving. First they had to learn how to manipulate the equations of relativity in such a way that they could even address the problem; then they went on to determine the types of waves that might be emitted by various astrophysical events. At the vanguard of this effort was Caltech's Kip Thorne.

Rai Weiss, admittedly no friend of theorists, makes an exception with Thorne. Introducing him one day at an MIT seminar, he noted that "Thorne is one of the most approachable theorists, a physicist who championed LIGO rather than sitting back and doing idle calculations." As a theorist, Thorne wears many hats. He has looked into the origins of classical space and time and explored the physics of a black hole. In the public's eye he is most notorious for his work on wormholes, hypothetical cosmic tunnels through hyperspace that provide shortcuts to both other reaches of the universe and other times. They sound like science fiction but are founded in genuine solutions to Einstein's field equations. Thorne started studying these weird entities in the mid-1980s because of a friend's request. Astronomer Carl Sagan, then working on his novel *Contact,* had asked Thorne whether there was a scientifically legitimate way for his characters to dart about the cosmos with ease. Thorne and his students came up with a solution that used wormholes. Their result, in a paper entitled "Wormholes, Time Machines, and the Weak Energy Condition," was published in the prestigious *Physical Review Letters* in 1988. But Thorne's most massive theoretical endeavor through most of his career, in collaboration with an armada of graduate students and postdocs, was carrying out the theoretical needs for LIGO—modeling potential cosmic sources of gravitational radiation and estimating the characteristics of their various waveforms.

Born in 1940, Thorne grew up in Logan, Utah, then a small college town of sixteen thousand. Although his parents were Mormons (their ancestors had moved west with Brigham Young), they didn't fit the group's typical conservative profile. Far more liberal, his father was an eminent soil chemist at Utah State University. His mother, who held a PhD in economics, initiated the women's studies program at Utah State and participated in anti–Vietnam War marches. The oldest of five children, Thorne caught the science bug early: "When I was eight, my mother took me to a lecture on the solar system given by a geology professor at the university. I was immediately fascinated. That was my first introduction to astronomy. Before that I wanted to be a snowplow driver. For a little boy growing up in a town in the mountains, where you have snow banks that are six or eight feet high, snowplow drivers are the most powerful people in the world."

By the time he was a teenager, a book by physicist George Gamow called *One, Two, Three, Infinity,* had hooked Thorne on relativity, and as a consequence geometry became a passion. He spent many summer hours working on problems in four dimensions. In high school, Thorne was a "cocky kid," as he puts it. Starting in the ninth grade, he sat in on college classes, including geology, world history, and mathematics. If bored in a high school class, he would just get up and leave. "They presumed I was just going off to the university," said Thorne.

Thorne's rebellion continued at Caltech, which he entered as an undergraduate in 1958. For three summers he worked at the Thiokol Chemical Corporation helping design rocket engines for the Minuteman missile program. When asked in his fourth summer to sign a loyalty oath, a legacy of the McCarthy era, he refused. He lost a prestigious National Science Foundation graduate school fellowship for the same reason (although he later received one when the loyalty oath requirement was finally dropped).

His choice of graduate school in 1962 was almost predetermined. Browsing through the physics journals, Thorne immediately recognized that the most interesting work in general relativity was being done at Princeton University, home of John Wheeler. In their first meeting, Thorne intently listened as Wheeler spoke for two hours outlining the outstanding problems of the time. Thorne chose to immerse himself in black hole physics, although that name wouldn't be officially used for five more years. Joe Weber was also a presence on campus, as he regularly shuttled between Maryland and Princeton to talk with Wheeler, Robert Dicke, and Freeman Dyson about the construction of his first bar detector. Thorne finished his

PhD in a speedy three years after writing a dissertation on hypothetical relativistic objects that were long, thin, and cylindrical. To his and everyone's surprise, he found that some of these unusual objects would be stable. Today, it's more than an academic exercise. Theorists wonder if the early universe, in the first fraction of a second of its existence, generated similar bodies, now called cosmic strings.

Thorne returned to Caltech as a postdoc in 1965, just as his skills in general relativity were most needed. Two years earlier quasars had been discovered, and some suspected black holes were involved. But others at Caltech, particularly William Fowler and Fred Hoyle, wondered whether supermassive stars were the source of a quasar's power. Fowler funneled students over to Thorne to help him look into such questions. When Fowler gave up his NSF grant in relativistic astrophysics after taking on new duties, Thorne essentially inherited the stipend, which allowed him to supervise and support a small army of PhD students and postdocs over the past few decades. As a result, Caltech eventually supplanted Princeton as the mecca for pursuing general relativity. Thorne became well known on campus in his younger days for his bohemian flair—shoulder-length hair, full beard, colorful shirts, and sandals.

Always Thorne made sure that his work touched bases with the real world. "There was a great richness of things to be done in bringing relativity in contact with the rest of physics," said Thorne. It was the legacy of Thorne's studies at Princeton, where he not only studied under Wheeler but also regularly dropped by Dicke's group to keep in touch with its experimental work. When Thorne was driving across the country with his family to make his move from Princeton back to Caltech, he recalled dropping by the University of Chicago to consult with the noted astrophysicist Subrahmanyan Chandrasekhar (a 1983 Nobel laureate for whom the Chandra X-Ray Observatory is named). They talked about neutron stars and gravitationally collapsed objects, which were then mere conjectures. "They seemed so far from any observation," said Thorne. "But Chandra expressed a confidence that neutron stars and what came to be known as black holes would, in some moderate number of years, be found. That had a big impact on me. I only wanted to work in areas that had an observational backup."

He didn't have long to wait. Pulsars were discovered in 1967, giving Thorne added confidence that his work on exotic objects would no longer be purely theoretical. But although neutron stars were gaining credence, black holes remained suspect. Many astronomers clung to the view that nature would somehow find a way to prevent a stellar core—one heavier than a neutron star—from collapsing into a singularity (in other words, a point). The upper limit on neutron stars is believed to be anywhere from one and a half to two and a half solar masses.

Stars lose mass as they age, and some figured at this time that they would always lose enough to drive them under the black hole limit. Thorne was working in an intellectual climate of great skepticism. "Not unlike the skepticism [first expressed] about gravitational-wave detection, whether waves are emitted sufficiently strong enough so that we will see them," he said. "The intellectual ambiance of that era was very much one that relativity was a beautiful subject but that it didn't have much to do with the real world. It was just a mathematical subject to be pursued for its own intellectual interest. This attitude pervaded physics from the mid-1930s into the 1970s. I found myself following Wheeler's footsteps as an advocate for these things being truly relevant to the astrophysical universe." Thorne's famous textbook *Gravitation,* published in 1973 with Misner and Wheeler, was to a large extent designed to promote the contact between relativity and the rest of physics. "It was in some sense a propaganda piece, as was my public lecturing. I was trying to convince the community that this was a field that was relevant to the rest of physics," said Thorne.

What ultimately turned things around were the observations arriving fast and furiously from new fields of astronomy, especially X-ray astronomy. The "final clincher," according to Thorne, was the close examination in 1971 of an exceptionally bright X-ray source located in the direction of the Cygnus constellation. Cygnus X-1's powerful X-rays come from a double star system consisting of a giant blue star and a dark, invisible companion whose measured mass of around fifteen solar masses strongly suggests it is a black hole. The X-rays are generated as matter, drawn away from the supergiant star, spirals inward toward the black hole.

Thorne carved out a special niche for himself in the general relativity community as he and his students began to scrutinize the stability of neutron stars and black holes. "All of this was done in the context that Weber was working on his experiment, and we were trying to understand the potential sources for his bar," said Thorne. "His experiments were very much on my mind."

At the time there were suspicions that if a black hole got spun up by accretion to a high speed—by stealing mass from a nearby companion—it would start vibrating and tear itself apart. That was one way to avoid having this ugly, awkward creature in the universe. But three of Thorne's students, Richard Price, William Press, and Saul Teukolsky, looked at the pulsations of a black hole and proved that if you perturb a black hole, the disturbance will quickly dampen as the energy radiates away as a set of gravity waves. In the end a black hole remains, and very much intact. Thorne showed that the same would be true for a neutron star. With these proofs in hand, Thorne had his students visit the math department to bring back new techniques for analyzing the radiation of gravity waves from stars and other systems. In test after test they and their colleagues at other universities came to the same conclusion: the formation of a black hole was inevitable if enough mass was around. Vibrations alone would not stop it. Even if one black hole were thrown at another black hole—one of the most cataclysmic cosmic events imaginable—what results is simply a bigger black hole that is perfectly stable.

In this way Thorne came to specialize in gravitational-wave physics. "I have an aversion to working in areas where other people are working because I prefer to do something that is unique. I don't like to be in the position of worrying that if I don't do something today a competitor will solve the problem tomorrow," he explained. Thorne was making a calculated bet. Most people in black hole research at the time didn't have high expectations that the technology would be good enough to see gravitational radiation anytime in the near future and so didn't care to examine its physics deeply. "But I thought there was a shot at it," he said, still comfortably attired in a loose cotton shirt but his once-long hair now gone.

It's a tricky business to determine just how much gravitational radiation is bathing the Earth daily. It strongly depends on the theoretical models for determining how much gravitational energy might escape an event, such as a supernova or a black hole collision. The models are complex, and gravity-wave detectors will have to collect a larger database of signals to make a reasonable assessment. Thorne has long maintained a chart of the possibilities, with the uppermost line marking off his "cherished belief" boundary, the strongest waves possible without violating any conventional beliefs about the nature of gravity. The sources that theoretical groups expect will be regularly recorded are many and assorted—from the violent collisions of celestial objects to the remnant wash of gravitational waves left by the Big Bang.

What gravity-wave researchers long considered their most exciting target—and the first they found—was the collision of two black holes, the biggest game of all in a gravity-wave hunt. Their first two 2015 sightings finally christened black holes as bona fide denizens of the universe. Up to that point, evidence for their existence had been circumstantial. For example, X-ray telescopes regularly pick up signals from remote orbiting bodies, which astronomers interpret as the high-energy radiation released right before a black hole permanently swallows the matter pulled off a companion star. Yet the black hole itself remains unseen. But two black holes orbiting each other reveal themselves when they eventually spiral into each other at a sizable fraction of the velocity of light, releasing an unmistakable set of gravitational waves that preserves a record of the fateful collision. It is a cosmic signature unique to black holes.

Picture two black holes slowly circling each other, like a pair of sumo wrestlers warily checking their opponent out in the ring. Tens of millions of years earlier these two black holes were simply stars, until they exhausted all their fusible fuel and collapsed to the most compact state imaginable. More than mere indentations in space-time, black holes are fathomless pits. No bits of light or matter can climb out of these deep gravitational abysses. That's why ordinary telescopes can't see them and theorists can do no more than imagine them. Only a gravity-wave telescope can detect them.

The sighting occurs only at the closing stage, after the two black holes have been slowly orbiting each other, perhaps over millions of years. During that time the pair would have been emitting a steady stream of very weak gravity waves, a wake that continually spreads outward along the canvas of space-time, like the spiraling pattern of a spinning pinwheel, as the black holes circle about. Gradually losing energy in this fashion, the two black holes relentlessly draw together as the years go by. And the closer they get, the faster they orbit each other.

In the final moments of this fateful dance, the gravity waves being emitted become strong enough to be detectable. Instruments on Earth register a whine, a series of waves that rapidly rise in pitch, like the sound of an ambulance siren that is swiftly approaching. These black holes should not be thought of as masses, Thorne has pointed out. Rather, they should be envisioned as whirling tornadoes of space-time, which are both dragging space-time around them as they orbit each other. "It's like two tornadoes encased in a third tornado, all coming together," he said. As the twirling black holes are about to meet, spiraling inward faster and faster at speeds close to that of light, the whine turns into a chirp, a birdlike trill that races up the scales in a matter of seconds, if the black holes weigh a few solar masses. If the holes are heavier (as was the case with GW150914), the chirp is far shorter. A cymbal-like crash heralds the final collision and merger. The two black holes become one. A ringdown, akin to the diminishing tone of a struck gong, follows as the new entity, a pit in space-time that swirls around like the fearsome tornado in *The Wizard of Oz*, wobbles a bit, and then settles down. "We have now embarked on an era of exploring phenomena in the universe that are made from warped space-time," said Thorne. "I like to call it the warped side of the universe."

The masses of the two black holes are determined from the duration of their observed coupling and the frequency of the gravity-wave signal. This was the one incontrovertible way that physicists finally clinched the existence of a black hole—nature's strangest star. The black holes gave themselves away by the melody of their gravity-wave song, the distinctive ripples of space-time curvature that they sent out into the universe as they merged. Gravity-wave astronomy will ultimately make black holes seem ordinary, noted Thorne.

All of this information would have been impossible to glean without another assist from the National Science Foundation. While LIGO was under construction, the agency also funded endeavors in gravitational theory. If and when they found a gravity wave, LIGO scientists needed improved mathematical models of their various sources in order to verify the signal. "Grand Challenges" were set up for teams of theorists and computer scientists to develop the needed codes, most notably for black holes colliding. "Through this extended effort," noted Richard Isaacson, "theorists worked just as hard as experimentalists. With a significant new budget of over one million dollars per year, perhaps the most lasting effect of this investment was the education and training of a new generation of young researchers working on simulations of solutions of the Einstein field equations . . . incorporating the full complexities of strong field gravity."

By the time the first-generation LIGO began operating, the very initial phases of two black holes spiraling in toward each other had been well modeled via post-Newtonian approximations. More troublesome were the later phases, which required numerical relativity. "Einstein's equations describe gravity via elegant but complicated nonlinear partial differential equations," noted Richard Matzner, director of the Center for Relativity at the University of Texas in Austin. Such equations cannot be solved by pencil and paper alone but rather require brute computation on the world's fastest and most powerful supercomputers. When Einstein's elegant equations are recast in this numerical mode, any one part can involve thousands of terms, which requires special software to handle. Because the solution requires a lot of computer power and memory, theorists tackled it in steps.

First, researchers handled the relatively simple case of two nonspinning black holes approaching each other head-on, meeting, and then merging. At each stage—from far encounter to near touching—the gravitational landscape changed, an evolution they captured step by step. But as far as nature goes, that's an unrealistic simulation. Like all other stellar objects, black holes spin. And when two black holes are involved, they will also be orbiting each other, which adds more complexity to the numerical simulations. The theorists' goal was to develop the tools needed first to model the process and then to optimize the calculations. This optimization depends on both how the calculations are set up and the power of the computer.

In the early 1990s scientists needed one hundred thousand hours of computer processing time (more than eleven years!) to perform an accurate three-dimensional simulation of two black holes spiraling into each other, which made the endeavor wishful thinking. But by the end of the decade, using algorithms optimized for parallel processing machines, that time was cut to a thousand hours, a much more doable prospect. But computer memory still needed to be increased mightily and new mathematical insights achieved before the full solution of a black hole–black hole collision was obtained. Coding the last few orbits and the final crash brought theorists to the cutting edge of general relativity's challenges.

Thorne had chaired NSF's advisory committee on the numerical relativity Grand Challenge and was increasingly dismayed by its progress. "They were not able to simulate a single orbit of two spinning black holes. They were developing methods that became important, but they were just plain hung up," he recalled. Supercomputers were crashing from the complexities. The numerical relativists were so far from a solution that Thorne made a bet with twenty-eight of his general relativity colleagues in 1995—"a bottle or bottles of wine, value not less than $100"—that LIGO would find a gravity wave *before* the numerical relativity community developed a code capable of handling two black holes merging. That was a worrisome prospect; it meant that LIGO might capture a wave without being able to interpret it. Thorne was so concerned about this possibility that he stepped away from his LIGO responsibilities around 2001 and initiated a collaboration with Cornell University physicist Saul Teukolsky to advance the computational and mathematical tools needed to predict the behavior of gravitational-wave sources. Soon other institutions in the United States, Canada, and Germany joined this endeavor, known as SXS for Simulating Extreme Spacetimes. Other theoretical groups around the world also began tackling the problem.

A huge breakthrough arrived in 2005 when Princeton physicist Frans Pretorius, then a postdoc at Caltech, decided to go in a different direction from his SXS colleagues. Using a combination of techniques that had not yet been tried, such as adopting a particular formulation of Einstein's equations with harmonic coordinates, he succeeded in simulating the first black hole binary merger—five orbits of two holes spiraling inward and then colliding—along with the gravitational waves emitted during the event. His brute-force approach eliminated some spurious effects that had plagued earlier attacks on the problem. From that point on, other numerical relativists, such as Joan Centrella at the Goddard Space Flight Center in Maryland and Manuela Campanelli at the Rochester Institute of Technology in New York, led teams that were able to develop additional techniques that moved the field forward speedily. In the following years computational theorists came to generate the necessary gravity wave templates for recognizing black hole mergers with a wide range of masses, spins, and orientations—

all before LIGO made its first detection. Thorne, thankfully and with no regrets, settled his bet.

Another collision that LIGO is almost guaranteed to hear is the resounding crash that occurs when two neutron stars, paired together in a binary system, spiral into each other as their orbital dance decays. In 1963, four years before the first neutron star was even discovered, physicist Freeman Dyson estimated the gravitational radiation expected from such a neutron star pair. "It would seem worthwhile to maintain a watch for events of this kind, using Weber's equipment or some suitable modification of it," he wrote at the time. It was a prescient thought. When their instruments reach the necessary sensitivity, gravity-wave astronomers suspect that these events will be the bread and butter of their trade. As noted earlier, the two compact balls of matter in the famous Hulse-Taylor binary now emit gravitational waves around 0.0001 hertz. Only in its last fifteen minutes of life, when the two neutron stars have finally drawn quite close to merge, will the waves sweep from 10 to 1,000 hertz, setting off gravity-wave detectors on Earth. But the Hulse-Taylor binary won't be colliding for about three hundred million years. "So we must reach outside the galaxy to give our graduate students a thesis," joked Thorne. The length of the LIGO arms was chosen, in fact, ultimately to see such events out to a billion light-years and so obtain a good population of sources.

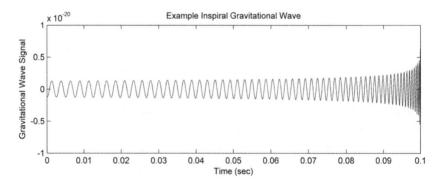

The type of waveform recorded by LIGO when neutron stars or black holes spiral into one another and collide. Each peak is a stretch; each trough a squeeze. The gravity waves emitted at first are fairly regular but the pace quickens right before the celestial bodies collide. *Courtesy of Amber L. Stuver.*

How often might this happen? That depends on both the statistics and the latest theoretical models. Theorists take the number of neutron star binaries known to exist in our own Milky Way galaxy and then extend that outward to encompass the vast volume of space that will be observable to LIGO. The event rate forecast for the first-generation LIGO was rather low, about ten per century at best. But when brought up to its top sensitivity, Advanced LIGO might see one or more a day.

These neutron star binaries will broadcast their own distinctive sets of whines and chirps. A neutron star is, in some sense, one big atomic nucleus. Only in this case the nucleus contains 10^{57} neutrons. "Somewhat larger than physicists are normally accustomed to," noted Thorne. Since using a particle accelerator to study such nuclear material is out of the question, neutron star binaries offer an alternative. More lightweight than black holes, a pair of neutron stars take longer to spiral in and merge than black holes in a comparable orbit. And they're also detectable at an earlier stage than black holes as they spiral inward, so the final recordable signal will last minutes instead of seconds. Gravity-wave telescopes will register a sinusoidal wave that sweeps to higher and higher frequencies as the two mountain-sized balls spiral into each other. Five to ten minutes before their lethal meeting, the two neutron stars are about five hundred miles apart and orbiting each other about ten times each second, at nearly a tenth the speed of light. In the final moments they are severely stretched by tidal forces and revolving around each other as much as a thousand times a second, dragging space-time around with them. These waveforms—the chirps—will contain within them a wealth of information for those who know how to look, such characteristics as the density and composition of the compact star's nuclear matter. (The waves generated as a neutron star plunges into a black hole will provide similar information.) As soon as they touch, the two stars are shredded to pieces, possibly releasing a burst of gamma rays.

Are those mysterious gamma-ray bursts in the sky the result of such a collision, as some suspect? Their distribution and intensity are in the right range. Roughly once a day a burst of gamma rays appears from some random direction on the celestial sky. On average, a flash lasts a few seconds, but the range extends from milliseconds to hours. Such bursts were first noticed by US Air Force Vela satellites, launched in the 1960s to monitor nuclear explosions on Earth. Evidence suggests that these bursts originate from far outside our galaxy. One of the brightest bursts to date was uncovered in 2009 by NASA's Fermi Gamma-Ray Space Telescope. It appeared to have the power of nine thousand ordinary supernovas. During the eruption it was as luminous as the rest of the universe. Soon after the burst was spotted, a ground-based detector at the European Southern Observatory in Chile recorded the afterglow of this mighty celestial fireball, which allowed astronomers to determine that the blast occurred more than twelve billion light-years away, nearly to the edge of the visible universe. This particular event went on far longer (twenty-three minutes) than the average and is believed to have been a massive star that ran out of nuclear fuel and collapsed into a black hole, with jets of material blasting outward at near the speed of light.

Two neutron stars colliding are expected to generate shorter gamma-ray bursts. But short bursts might also involve black hole collisions, a white dwarf star spiraling into a black hole, or every other possible combination of a white dwarf, neutron star, or black hole merging. A firm gravity-wave detection that coincides in both time and location with a gamma-ray burst will help sort it all out. (Intensely compact stars—such as white dwarfs, neutron stars, and black holes—are the best gravity-wave emitters. A regular star, like our Sun, would be too soft to stay intact as it approaches a compact star. Its matter would be stretched out by tidal forces before colliding, thus emitting weaker gravity waves that are harder to observe.)

But what happens *after* a neutron-star collision? The remnants might coalesce into a new, more massive neutron star, if it's rotating particularly fast. Or if heavy enough, they might condense to utter invisibility, forging a black hole. Only a gravity-wave telescope is able to reveal the final outcome. But once these signals are regularly detected, they would be a boon to cosmologists. Current measurements of distance rely on such yardsticks as the luminosity of stars and supernovas or the apparent size of galaxies. Astronomers continue to quibble over the interpretation of those standard candles. But by knowing the amount of gravitational energy emitted by inspiraling neutron star (and some black hole) pairs about to collide and comparing these estimates with the strength of the waves when they arrive on Earth, astronomers can calculate how far the waves had to travel to reach our planet. This, in turn, provides a measuring tape directly out to the events, without the worry of the intervening steps that can afflict other methods. Gravity-wave astronomers will then be able to use all these events—out to the farthest reaches of the cosmos—to map the evolving geometry of the universe over time. The information gathered from all those binary mergers, for example, will serve as a double-check that the universe's expansion is indeed accelerating, as other astronomical observations suggest.

There is another type of signal in the gravity-wave sky as well, although it will be less frequently recorded because of its weakness compared to binary-star collisions. A solitary tsunami of a wave will hit our shores every once in a while, generated at the very moment a star explodes in our galactic neighborhood as a brilliant supernova, its core crumpling up to form either a dense neutron star or a black hole. This core collapse triggers a shock wave that blows off the star's remaining, outer mantle of gases, which we see as a supernova. Astronomers routinely spot these explosions in other galaxies with the aid of telescopes. For laypersons using only their eyes, it requires a little more patience. The last supernova visible to the naked eye occurred in 1987. The explosion was sighted in the southern sky within the Large Magellanic Cloud, a satellite companion to the Milky Way. The last visible supernova in our galactic neighborhood before that was spotted by Kepler in 1604.

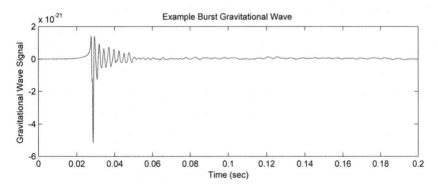

Brief bursts of gravity waves are expected from short-duration events, such as stars exploding as supernovas. Here we see a sudden and big wave that quickly diminishes. *Courtesy of Amber L. Stuver.*

Sighting a supernova with gravity-wave detectors is not assured, though. A lot will depend on the explosion's dynamics. If the collapse of the remnant core is perfectly smooth and symmetrical, gravity-wave astronomers will not hear even a whimper; gravity waves emitted symmetrically cancel each other out, much the way out-of-phase light waves do. At the same time that one part of the wave is causing space to stretch, another part is causing it to contract; the net result is no change at all. Gravity waves would be emitted only if the collapse is a messy affair, with the newborn neutron star squishing down like a pancake and then stretching out before settling down. As a result, a gravity signal will be sent out. If the core is spinning madly at the end of its life, it could even flatten and be turned into a barlike configuration, spinning end over end like a football. In that case the collapsing core would send out very strong waves. The Advanced LIGO detector might see several a year, which would make it one of LIGO's more dependable sources.

There is some evidence that supernova explosions can be imbalanced. Astronomers have seen individual pulsars speeding through the galaxy at velocities greater than a hundred miles per second. It is suspected that these pulsars got shot out by an asymmetric explosion, an extra kick on one side more than the other. Supernovas occur within our galaxy only two or three times every century on average, but their signal strength under these circumstances would be spectacular. It is estimated that the supernova seen in 1987 had a strength that initial LIGO was capable of detecting.

All the while, playing in the background of this gravity-wave symphony, could be a steady hum. When a neutron star forms or if its outer crust (ten billion times stronger than steel) later undergoes a strain, for instance, it might briefly vibrate and develop a bump on its surface, a sort of a "mountain" that grows a few inches and freezes into place. And as the neutron star feverishly whirls around, this deformation, jutting out like a finger, would send out a continuous gravity wave as it "scrapes" the space around it. For a neutron star rotating once every thousandth of a second, its equator ends up spinning at about 20 percent of the speed of light. Deformed neutron stars could serve as gravity-wave lighthouses scattered over the heavens, each sending out a wave until its lump smoothes out over time. (If the neutrons—each composed of three quarks—have degenerated into the quarks themselves,

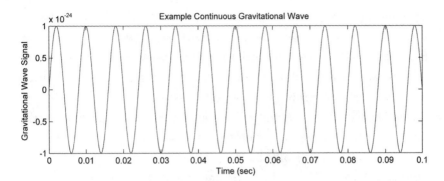

A neutron star with a "mountain" (deformation) on its surface is expected to emit a weak, continuous gravity wave as it swiftly rotates about its axis. *Courtesy of Amber L. Stuver.*

forming a hypothetical quark star, then the mountains might be around ten meters, or thirty-three feet, in height. Built out of more exotic material, a quark star is denser and so could sustain higher deformations.)

The only interruption to this continuous signal might be an occasional gravity-wave burp, released whenever the neutron star undergoes a starquake. This could happen when the outer crust of the neutron star slips at times over its superfluid core. A neutron star's continuous signal would be extremely weak, too weak to see right away. In this case the interferometer has to gather the data for a long period and add them up in order to have the signal emerge from the background noise. No gravitational radiation has yet been detected from such sources, but from the searches done so far, LIGO-Virgo collaborators were able to set some upper limits for two, relatively nearby pulsars—the neutron stars within the Crab and Vela nebulas. They are fairly sure that the Crab pulsar has no hills greater than three feet, while the Vela pulsar has none higher than thirty-three feet (otherwise, current detectors would have seen them by now).

Theorists have also been excited by a calculation that suggests that a neutron star's dense nuclear matter might actually slosh around soon after the star forms, fed by the star's rapid spin. Such oscillations occur in our own oceans, creating circulation patterns. In the case of a neutron star, gravity waves are generated. More interesting, the gravity waves increase the sloshing, which in turn increases the production of gravity waves. How large does the sloshing get? Theorists don't know but suspect that other forces, such as friction and magnetic fields, step in at some point to cut it off. Until that happens, though, the newborn neutron star might release a unique gravity-wave cry as it cools and settles down, enabling astronomers to investigate the star's interior from many light-years away.

And beneath the chirps and pops emanating from the gravity-wave sky, there could be an underlying murmur—constant, unvarying, and as delicate as a whisper. This gentle buzz would be the faint reverberation of our universe's creation, its remnant thunder echoing down the passages of time, similar to the residual microwave heat already detected from the Big Bang. The cosmic microwave background, however, only says something about the universe's condition nearly half a million years after the Big Bang. That's when the primordial fog lifted and the universe became transparent. By then the universe had cooled down enough for neutral atoms to form, at last allowing radiation to travel unimpeded. These light waves, mostly in the optical and infrared regions of the spectrum by that point, were gradually stretched with the universe's expansion until today they are detected as a vast sea of microwaves. Before that decisive moment in our universe's history, the primordial fireball was a murky soup—a jumble of protons, simple nuclei, electrons, neutrinos, and photons of electromagnetic radiation, all intermixed. Even if astronomers were someday able to peer back to this epoch, they wouldn't see much, for the cosmic plasma was quite opaque, just as the Sun's hot outer layers prevent us from gazing into its nuclear core. This so-called fireball is an impenetrable barrier to our electromagnetic view. Primordial gravity waves, however, cut right through that fog.

A portion of those waves would be fossils from the very instant of creation, tiny jiggles in space pumped up by an explosive burst of expansion that took place a scant 10^{-43} second into the universe's birth. No other signal survives from that era. These relic waves would bring us the closest ever to our origins. At the same time they might tell us how fast the universe expanded over the eons. These primordial waves could be detected in a roundabout way. Teams of radio astronomers operating very sensitive instruments in such places as the South Pole and Chile are on the lookout for the imprint of these gravity waves upon the cosmic microwave background—the now-weak afterglow of the Big Bang. The gravity waves impose a faint spiraling pattern into the polarization of the microwave signal. But capturing the relic gravity waves *directly* would bring us the closest ever to our origins, perhaps verifying that the universe emerged as a sort of quantum fluctuation out of nothingness. The gravity-wave background is too weak for current ground-based interferometers to detect, but space interferometers (possibly a second generation) will have a chance.

Moreover, there could be other gravity-wave events at later moments in the universe's birth, when the universe experienced an abrupt change in its environment—a sudden shift or phase transition between two different states, like liquid water turning into something far different, ice. As the universe expanded, it too likely underwent some kind of phase transitions as it cooled. Additional gravity waves might have been emitted in the process. During the first searing flash of creation, for example, theorists believe that the four forces of nature were united. But then, as the universe coursed outward, the individual forces swiftly broke away from each other to form separate identities—first gravity, then the strong nuclear force, and last, at about a trillionth of a second after the Big Bang, the electroweak force divided into its two separate components, electromagnetism and the weak nuclear force. Slow this last transition down and it would appear like water droplets forming in a cloud: within certain regions—the so-called

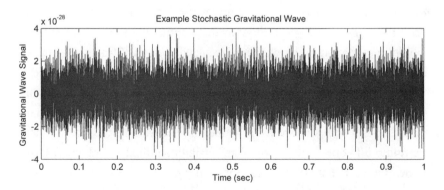

Some sources will appear as a very weak, continual "buzz." One example would be the gravitational waves emitted in the earliest moments of the Big Bang and now spread out into every part of the celestial sky with the expanding universe. *Courtesy of Amber L. Stuver.*

droplets—you would have the electromagnetic force, outside of which it doesn't exist. "And those droplets expand at the speed of light and collide with one another," explains Thorne, "producing gravitational waves." Thus, with a space interferometer, physicists have a chance at watching the birth of the electromagnetic force.

The rapid transitions could also have given rise to objects that retain features of the earlier higher-energy state. There is still the chance that strange new celestial creatures could greet us as gravity-wave astronomers examine this realm. Not until astronomers scanned the heavens with radio telescopes did they discover pulsars and quasars; neutron stars had been contemplated, but not as pulsing radio beacons; quasars were never even fantasized. What else might be skulking about in the darkness of space as yet unseen? Pulsars and quasars may turn out to be commonplace in comparison to the exotic astrophysical events that gravity-wave astronomy reveals. Some theorists already wonder whether there might be highly energetic, so-called defects that were generated as the cosmos cooled down over its first second of existence. Their names only hint at their strangeness—pointlike monopoles, one-dimensional cosmic strings, and domain walls.

Cosmic strings are one of the more interesting defects hypothesized. One might think of them as extremely thin tubes of space-time, skinnier than a subatomic particle, in which the energetic conditions of the primeval fireball still prevail. Any strings that survived to this day would be either exceptionally long (spanning the entire width of the universe) or bent back on themselves, creating closed loops that continually lose mass-energy by vibrating at velocities approaching the speed of light. Gradually losing energy, they'd finally rocket away in their death throes, disappearing with a final blast of gravitational radiation. If these potent strings truly exist, astronomers may not be too desirous to observe them close-up. While such a string, so exceedingly thin, could actually whiz through your body without bumping into one atom, its peculiar gravitational field would wreak havoc nonetheless: if this string sliced through you, your head and feet would proceed to rush toward one another at ten thousand miles per hour. Because of the tremendous tension in a string, it would wiggle around like a rubber band and crack like a whip, producing lots of gravity waves. Such gravitational radiation emanating throughout the universe could very well affect the timing of radio pulsars if the gravity waves pass by any of those neutron stars (astronomers are checking; more on that in the next chapter).

Perhaps more exciting is the prospect of encountering the unanticipated. For one, the exact form of a gravity-wave signal might disagree with the predictions of general relativity. It hasn't so far, but if it ever does, that might indicate that Einstein's equations have to be amended when dealing with sources that involve a horrifically strong gravitational field. In that case, gravity-wave astronomy's findings would usher in a new physics of gravity, akin to the way Einstein supplanted Newton. It might even provide clues as to how theorists could forge a theory of everything that unites general relativity (which rules the macrocosm) with quantum mechanics (the laws of the microcosm).

The Unending Symphony

WITH TEST AFTER TEST CURRENTLY SHOWING COMPLETE agreement with Einstein's general theory of relativity, it's natural to ask why physicists put such effort into designing ever more sophisticated experiments. Because, says general relativist Clifford Will, "every test of the theory is potentially a deadly test." Gravity is the force that rules the universe. To understand its workings, to the finest degree, is to understand the very nature of our celestial home. Any deviation, any surprising signal, surely offers clues to an even deeper understanding of the cosmic design. Consequently, gravity-wave astronomers are not satisfied with Earth-based interferometers alone. Interferometers on the ground are limited in the range of frequencies they can detect, which is unfortunate because many interesting events originating from strong astrophysical sources are expected to occur in the very low frequencies, from a millionth of a hertz to 1 hertz. To examine those regions, gravity-wave astronomers must venture into space.

In some ways, gravity-wave astronomers have already established a presence in space with the many spacecraft that have sped or at this moment are speeding through the solar system on their journeys to various planets. In each case there are two masses: one is the spacecraft; the other is the Earth. From Earth a signal with a very precise frequency is transmitted to the spacecraft, and a transponder aboard the spacecraft sends it back. If a gravity wave passes by and jiggles the Earth, it will shift the frequency of the signal it is sending out ever so slightly. Whenever the gravity wave hits the spacecraft, it will cause a similar frequency change in the spacecraft's re-turning signal. All in all (as soon as all other sources of noise are subtracted out), what's left is a distinctive set of pulses, the imprint of the radio wave being intermittently altered by the gravity wave. The entire Earth-spacecraft system acts in some way like an interferometer with a single arm hundreds of millions of miles in length.

Planetary scientists have already monitored the communications of the Viking, Voyager, Pioneer 10 and 11, Ulysses, and Galileo spacecraft. The chances were always small that anything would be detected; fluctuations in the solar wind can alter the frequency of a radio beam. But if a particularly big gravity wave had passed by during those missions, the opportunity was there. The gravity waves capable of being found by this method would be extremely long. From peak to peak one full wave could stretch from here to the Sun or even farther. Such a lengthy undulation might have been sent out by two su-permassive black holes in close orbit around each other in some far-off galaxy.

One of the biggest tests of this kind was carried out in the spring of 1993. At that time three separate spacecraft were speeding from Earth in different directions, which offered the perfect opportunity to test for gravity waves. NASA's Mars Observer (before it malfunctioned) was headed toward the red planet, the Galileo probe was trekking toward Jupiter, and the European Space Agency's Ulysses was journeying toward the Sun. Between March 21 and April 11 of that year, radio signals were simultaneously beamed toward the three spacecraft using a network of radio antennas situated around the globe (NASA's Deep Space Network). Once received, each craft amplified the signal and sent it back to Earth. The American and Italian planetary scientists running the test figured that if a gravity wave had passed by, the spacecraft would have gently rocked, like buoys bobbing on the ocean of space-time. That is why the frequency of the radio transmission would have shifted, ever so minutely. A change from just one spacecraft might have been spurious, the result of a local disturbance. But if a frequency shift had been detected in the signals from all three probes around the same time, the evidence would be more convincing. Superaccurate atomic clocks, timepieces capable of discerning a change in frequency of a few parts in a million billion, were used to monitor the potential shifts. Nothing resem-bling a gravity wave was seen, but that was not too surprising. It was a long shot at best. Spacecraft are still buffeted by the solar wind as they cruise through space, and the Earth's turbulent atmosphere can introduce some radio noise as well.

For the 1993 test the frequencies used to communicate with the space-craft were between 2.3 and 8.4 gigahertz. But the use of an even higher frequency cuts down on interference from the plasma in the Earth's iono-sphere as well as the solar wind. That was the case with the Cassini mis-sion, a spacecraft launched in October 1997 to study Jupiter and Saturn. "We were hitchhikers before," said John Armstrong of NASA's Jet Propul-sion Laboratory (JPL). "But now on Cassini, there was specific hardware on board to carry out these gravity-wave experiments." Cassini used a frequen-cy of 32 gigahertz, four times higher than in previous space tests. There were three opportunities to search for gravity waves between Cassini and Earth: forty-day periods from December to January in 2001, 2002, and 2003. When such tests were first done in the 1980s with Pioneer 10 and 11,

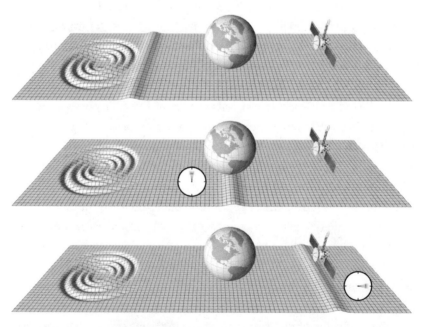

A gravity wave arrives from a cosmic event. One way to detect the wave is to compare how the wave affects the speed of a clock on Earth (*middle*), then one on a spacecraft situated in the far solar system (*bottom*).

the potential gravity-wave strains that the Earth-spacecraft system could measure were about 10^{-13}. Cassini had the capability to do a thousand times better, down to a strain of about 10^{-16}, largely owing to its use of the higher communication frequency. That's also ten to thirty times better than the 1993 experiment. The Cassini experiment, in fact, was about the best that could be done at the time via this method of testing. Another way would be to send synchronized atomic clocks directly into space, separated by about a hundred million miles or more, and watch time subtly change (be gained or lost by one part in a million trillion) as a gravity wave passes by. Needed are more precise atomic clocks, but until then, astronomers came up with another clever strategy—pulsar timing.

This scheme is based on those well-studied astronomical objects. These speedily twirling neutron stars are the most exquisite timepieces in the universe. By monitoring the steady radio beeps arriving from an array of particularly fast pulsars, situated around our galactic neighborhood, astronomers are on the lookout for slight changes in the pulsing caused by an extremely low-frequency gravity wave (10^{-9} to 10^{-6} hertz) passing between the pulsar and radio telescopes on Earth. Converting frequency to wavelength, this means that *one wave alone* would stretch anywhere from two hundred billion miles to two hundred trillion miles (three hundred billion kilometers to three hundred trillion kilometers) from peak to peak. Put another way, it would take anywhere from days to years for just one wave to pass by the Earth. Such tremendously long waves are emitted as two supermassive black holes slowly orbit each other, a binary that can form when two galaxies (each with a monster black hole in its center) merge. Such a binary will likely arise when the Milky Way collides and blends with our neighbor the Andromeda galaxy some four billion years from now.

But unlike LIGO's first sighting, a pulsar-timing discovery will arrive, not as a distinct signal, but as a result that emerges after many pulses are gathered and analyzed. "The smoking gun," said NASA Jet Propulsion Laboratory researcher Stephen Taylor, "will be seeing the same pattern of deviations in all of [the pulsars]."

"We're like a spider at the center of a web," added Michele Vallisneri, Taylor's colleague at JPL. "The more strands we have in our web of pulsars, the more likely we are to sense when a gravitational wave passes by."

NANOGrav, the North American Nanohertz Observatory for Gravitational Waves, has already been monitoring more than fifty pulsars for about a decade. NANOGrav astronomers are hoping for their first results within a few years. To carry out this drawn-out search, they use two of the most sensitive radio telescopes on Earth—the Green Bank Telescope in West Virginia and the Arecibo Observatory in Puerto Rico. They also collaborate with the International Pulsar Timing Array, a consortium that includes pulsar timing groups in both Europe and Australia, all pooling their resources with the aim of detecting very-low-frequency gravity waves from supermassive black hole binaries.

But when such orbiting supermassive black holes draw closer over time, the gravitational waves they send out get shorter and shorter—eventually too short for the pulsar timing arrays to pick them out. To see the two holes actually merge will require sending an entire interferometer into space. And such a scheme was imagined very early in the history of gravitational-wave astronomy.

The topic came up in 1975 during a meeting that Rai Weiss convened for his NASA committee looking into gravitational space physics. A group from NASA's research facility in Huntsville, Alabama, made a specific proposal to send a huge laser interferometer into space. This team suggested that two aluminum trusses, each about a kilometer in length, be manufactured in space and then joined in the form of a cross. The test masses would be suspended from this structure. It was a time when scientists were eagerly discussing the prospects of sending a variety of instruments into space. After the meeting, over dinner at a seafood restaurant, Weiss asked Peter Bender whether such a gravity-wave antenna sounded like a sensible idea. During the ensuing conversation, they began to think of using separate spacecraft rather than a rigid structure, which allowed the mirrors to be many miles apart.

Bender was a member of Weiss's committee because of his previous work on a lunar ranging experiment. Trained as an atomic physicist by Robert Dicke at Princeton in the 1950s, Bender went on to work at the National Bureau of Standards and later with the Joint Institute for Laboratory Astrophysics at the University of Colorado in Boulder. His work on precision distance measurements commenced in 1962, when he began a long-term collaboration with James Faller, another student of Dicke's who had newly arrived at JILA as a postdoc. Faller was eager to convince NASA to deposit a package of reflectors on the surface of the Moon during an unmanned lander mission in order to reflect a laser beam sent from Earth. By measuring the travel time as the beam bounced between the Earth and the Moon, researchers could determine the lunar orbit very accurately, as well as learn more about the lunar interior. Later it came to be recognized that the reflectors could also be used to carry out tests of general relativity, particularly whether the Moon and the Earth differed in their acceleration toward the Sun. By 1965 Faller's idea, with others brought on, turned into the Lunar Ranging Experiment (LURE). Sheer luck allowed the LURE team to piggyback its project onto the Apollo 11 mission in 1969, the first manned landing on the Moon. Worried that the Apollo 11 astronauts would not have time to carry out all their planned experiments, NASA began looking for projects that didn't require much setup time. The LURE proposal was perfect. The astronauts simply had to set the reflector package on the lunar surface and adjust the equipment to the proper angle to reflect a laser beam from Earth. Additional reflectors were set on the Moon during other lunar missions.

LURE turned out to be a very hardy experiment. Indeed, the system is still working today, the last experiment to continue operating from the Apollo program. The passive instruments just keep on reflecting. Although there has been degradation from heating and dust, the reflectors remain functional; laser tests with them continue to be carried out from the Apache Point Observatory in New Mexico. "It's tested Einstein's Strong Equivalence Principle to a part in a thousand. If you look at tests of relativity, that's one of the major ones that's been done," noted Bender. It shows that gravity accelerates objects equally, regardless of their mass or energy. The Earth and the Moon have been found to accelerate toward the Sun at the very same rate. It's the space-age extension of the Leaning Tower experiment.

That was Bender's first taste of experimental relativity. His dinner with Rai Weiss in 1975 launched a far more serious involvement. Their conversation about the Huntsville proposal for a massive space-based interferometer soon expanded in subsequent weeks to include both Faller and Ron Drever. "We had recognized that the ends of the interferometer should be separated as far as possible—that you didn't need a fixed structure," explained Bender.

"We finally got up our nerve to talk about a thousand kilometers or so, without realizing that others had mentioned such a distance with separate spacecraft considerably earlier."

In the early 1970s a number of scientists, including Forward and his colleagues at Hughes, theorists William Press and Kip Thorne, as well as Vladimir Braginsky in the Soviet Union, remarked in print on the possibility of taking interferometers into space as separated spacecraft, not long after Weiss had put his first laser interferometer design down on paper. "Then Ron said, 'Why stop there?' You just gain by making it longer. We ultimately concluded that a distance of a million kilometers was reasonable to think about," recalled Bender. Here was the kernel of an idea that has evolved and gestated for decades.

The idea of launching a gravity-wave detector into space at first seemed quite fanciful. It was partly hampered by the impression that laser interferometer technology was far from being ready. At the time, stable lasers had relatively short lifetimes and low power. With no improvements, the arms of a space interferometer spanning some one million kilometers (about 621,000 miles) would have had to match within ten meters, or just under thirty-three feet. Such fine-tuning would have required frequent adjustments to the spacecrafts' positions, interruptions that would make measurements far more difficult. But once the idea was on the table, interested researchers began thinking of ways around these obstacles. Faller, who gave the first public talk about the concept in 1981, came up with a way to subtract out the laser noise. Bender, using his expertise in celestial mechanics from the lunar ranging experiment, worked out the best orbit—a heliocentric path about the Sun.

Faller and Bender kept the idea alive, aided by modest grants from both NASA and the National Bureau of Standards (now the National Institute of Standards and Technology). This money enabled R. Tucker Stebbins, who had been involved in experimental relativity since his undergraduate days, to join Bender and Faller to put a more sustained effort into the proposal. They even gave their proposed space detector a name: LAGOS, for Laser Gravitational-Wave Observatory in Space. By 1989 LAGOS received high marks from a NASA committee looking into possible space ventures for astronomy after the completion of its Great Observatories program. But

From left to right, James Faller, Peter Bender, and R. Tucker
Stebbins initiated research on a space-based laser
interferometer at the University of Colorado's Joint Institute
for Laboratory Astrophysics. *Ken Abbott;* © *University of
Colorado.*

when NASA's advanced research funding got cut a few years later, interest
turned tepid. "There was even a joking comment about putting us out of
our misery," said Stebbins.

The idea revived, however, when a number of American space interferometer veterans joined forces with a larger group of European gravity-wave
specialists, including GEO600's Karsten Danzmann, Jim Hough, and Bernard Schutz. Their formal proposal for a mission, newly dubbed LISA for
Laser Interferometer Space Antenna, was presented to the European Space
Agency in 1993. The project was eventually accepted as an ESA "cornerstone mission." The ESA decision, though, was made with the expectation

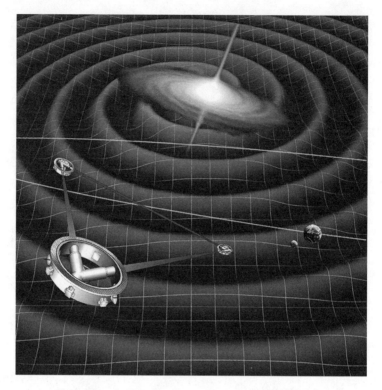

An illustration showing the three LISA spacecraft in formation,
awaiting a gravity wave. *Courtesy of ESA.*

that NASA would come on board as an equal collaborator. Before ESA
stepped in, the LISA project had been a labor of love, with various participants working on it in their spare time and using their own discretionary
funds to finance the initial studies. But once ESA provided seed money,
"many nooks and crannies of the design were looked at for the first time,"
said Stebbins.

In 2011, NASA announced that it could not participate because of budget
cuts, but ESA is continuing to work on the idea in a reduced form. The
design for this space interferometer is not yet firm, but it will likely resemble Bender and Faller's original idea. That version envisioned three
spacecraft flying in a triangular formation, with the center of the triangle
tracing Earth's orbit. In this way, the entire system perpetually follows the
Earth like a faithful companion at a distance of about thirty million miles.

Being so far out in space, there will be no seismic disturbances. The test masses, likely polished gold-platinum cubes 1.8 inches (47 millimeters) wide, will be free-falling along a space-time pathway carved out by the Sun. The orbits were chosen so that each spacecraft will have solar illumination from a constant direction, which will help maintain a stable thermal environment. Once properly inserted into orbit, the three spacecraft should fly in formation with little adjustment, with just a very weak and steady thrust to counter solar radiation pressure, which would otherwise push the spacecraft like wind powering a sailboat.

The three spacecraft will be separated by a sizable distance, possibly six hundred thousand miles or more. Each will be carrying lasers and test masses arranged in the form of a Y, so that each spacecraft can be aimed at the others. This will allow laser light to be continually transmitted and received along either two arms or all three. Because of the long distances involved, LISA requires a different approach to interferometry. As the laser light travels the long distance from spacecraft to spacecraft, the beam will get wider and wider, eventually spreading out some dozen miles. If the originating laser has 2 watts of power, for example, that will diminish to less than a billionth of a watt at the point of arrival. So the signal cannot simply be reflected back. Rather, the beam must first be amplified—the signal boosted—with the onboard laser before being sent back along the arm. If the beam were simply reflected, only one photon would make it back every three days, which would make measurement impossible. Such an amplification scheme is already used in tracking spacecraft, but with radio waves instead of laser beams.

Other engineering needs for LISA require sensitivities and specifications now being tested. Most important, there must be a means for maintaining a drag-free environment so that all forces upon the test masses (besides solar and planetary gravitational forces) are nearly completely eliminated. With each test mass being in free fall, isolated as if it were a separate body floating freely in space, the walls housing each mass must never touch it. Over any one second each chamber must not move in relation to its test mass by more than several nanometers. That's an extremely small distance, a span just a few atoms thick. The technology to conduct such maneuvers has been used on other satellites, but not to the fine performance needed by LISA. The adjustments will have to be made with microthrusters. "Essentially we need the most gentle rocket you can imagine," said Stebbins. Moreover, they have to worry about such things as the occasional micrometeoroid strike, which would add a sizable push to the spacecraft. If a dust particle, just a few thousandths of an inch wide, hit one of the spacecraft, it would disturb the test mass enough that the microthrusters would have to turn on to maximum thrust temporarily to correct for the shift.

ESA launched LISA Pathfinder into space at the end of 2015 with a package onboard that tested many of the new technologies needed for LISA. In this mission researchers learned how to maintain the test masses in free fall and measured their relative motions with record-making accuracy. In fact, said Danzmann, Pathfinder's performance was so much better than expected that he sees no overwhelming problem in going ahead with plans for LISA, which could fly sometime in the 2030s if fully funded.

LISA would be a complement rather than a competitor to the ground-based interferometers. That's because it would be receiving very-low-frequency gravity waves, from 0.0001 hertz to 0.1 hertz, well below the band accessible by LIGO, Virgo, or other ground-based detectors. As with pulsar timing, LISA will be detecting the huge, long swells in the ocean of space-time, while LIGO and its related kin observe the finer ripples. Each type of wave is generated by either a different astronomical source or a different moment of an event. LIGO and Virgo, for example, are best tuned to see black hole–black hole binaries when each hole weighs up to a few dozen solar masses. LISA, in contrast, is designed to spot black hole systems in the range of one hundred to one hundred million solar masses.

Neutron star binaries will be visible to both types of interferometers but at different times. LISA will see them as two stars orbiting each other, thousands of years before they collide. LIGO and Virgo observe them right before the collision, as the stars rapidly spiral inward and the gravity waves sweep to higher and higher frequencies. Thus, instruments are needed both in space and on the ground to cover the entire spectrum of waves likely to journey through the heavens. A typical wave for LISA, though, will be rather lumbering, taking perhaps a thousand seconds (sixteen and a half minutes) or more for just one wave to pass by from peak to peak. This is the type of astronomy for those with a patient temperament. Consequently, LISA will ultimately gather fewer data than detectors on the ground. LISA's data rate will be a few thousand bits a second, versus eighty million bits a second in LIGO. One flash drive will be able to store the data from the entire LISA mission.

There is one reason—and one reason only—that supporters have stuck with the idea of a space interferometer through thick and thin financial times (and why the Chinese are also looking into building their own or joining up with LISA). If the technology works, space interferometer observers are guaranteed to see something. LISA or its equivalent will be overwhelmed by sources, inundated by a fairly noisy background. Galactic binaries, such as the myriad number of white dwarf stars orbiting one another throughout our galaxy, will be broadcasting an unremitting cacophony of waves discernible from space. These sources are considered so surefire that "if [a space interferometer] would not detect the gravitational waves from known binaries with the intensity . . . predicted by General Relativity," reported a LISA study team, "it will shake the very foundations of gravitational physics."

The expectation is that, once these binary signals from our galaxy are examined and understood, they can be subtracted from the space interferometer data. "What you're left with is a record that should have the extragalactic information in it," said Bender. That would be the signals from massive black holes in faraway galaxies. Space interferometers would be vital instruments for studying such supermassive black holes in distant galaxies. These would be gravity-wave astronomy's most powerful sources. Their examination, across the breadth of space-time, is one of LISA's prime objectives. One of the best routes for seeing the signals from supermassive black holes, gargantuan objects containing the mass of millions of suns, will be catching two galaxies in the act of merging. During this process, the black holes in each galactic core would eventually coalesce to form an even bigger hole, resulting in a tremendous burst of gravitational radiation. LISA should be capable of seeing the entire last year of the inspiraling holes before their fateful collision. University of Michigan astronomer Douglas Richstone, a longtime supermassive black hole hunter, calls this "the brightest object in the sky that has not yet been seen." LISA has the potential to detect ten such events each year. It would be sensitive enough to register waves arriving from as far out as twelve billion light-years, across nearly the entire visible universe.

The chances are quite high that a space interferometer will see something, for evidence has been emerging in recent years that all galaxies with a spheroidal bulge, such as elliptical galaxies and most spiraling galaxies, harbor a supermassive black hole in their centers. The larger the bulge, the larger the black hole. Our own Milky Way galaxy harbors a black hole of around four million solar masses smack dab in its center. Evidence is mounting that these holes are "quasar fossils," the engines that once allowed each galaxy when it was young to blaze away with the brilliance of up to a trillion suns. It suggests that the formation of a galaxy and the growth of its central black hole are intimately linked in some as yet undetermined way. Perhaps a modest black hole forms first, serving as the gravitational seed around which a galaxy forms. Over time such a black hole would consume much of the young galaxy's rich supply of stars and gas, bulking itself up. The black hole becomes supermassive. Or perhaps the early universe generated a bevy of smaller black holes, each in an individual galactic building block. These separate pieces could then have eventually merged to form a full-blown galaxy, while the black holes coalesced to form a gargantuan black hole in its center. Another possibility: during the turbulent chaos of a new galaxy's formation, huge amounts of gas accumulated in the galactic center, reaching such high densities that the gas didn't turn into stars but directly collapsed into a massive black hole.

Whatever the case, gigantic black holes appear to be the natural result of galactic evolution and serve to power the fireworks display that announces each galaxy's birth. This happens because the massive black hole gravitationally attracts any matter lurking near it and never lets it go. But before this material is permanently captured, it gathers into a swirling accretion disk that surrounds the black hole and radiates intensely. At the same time the hole also spins, like a giant electromagnetic dynamo, producing two jets of subatomic particles that shoot away in opposite directions from each pole at near the speed of light. All of this activity occurs as long as there is enough fuel nearby—stars, dust, and gas—to feed the dark monster in the middle.

Today, quiescent holes can be reignited when galaxies collide. And there are hints in the celestial sky that mergers are in progress. The radio galaxy 3C75, for example, has a set of curving radio jets that resemble a giant water sprinkler at work. These jets appear to be emanating from two nuclei, each possibly an immense black hole. The long spouts get twisted as the two black holes orbit each other. These particular black holes won't be merging for many, many years. But when they at last approach each other, they will emit a distinctive gravity-wave signal. The frequency will start low and sweep to higher and higher frequencies as the year progresses and the holes get closer. It will be gravity-wave astronomy's ultimate payoff: conclusive proof that black holes are indeed the engines of a galaxy's central activity. A space interferometer could serve as an early warning system. If it can precisely pinpoint the location of an active inspiraling, optical, X-ray, and gamma-ray detectors could be trained on that region to record the final collision.

Almost as interesting will be watching smaller objects—neutron stars, smaller black holes, white dwarfs, and ordinary stars—falling into a supermassive black hole at a galaxy's center. Because stars are so plentiful, this could occur fairly frequently. The final orbits of the doomed stars would be intricate and complex. In some cases, the orbital decay could go on for seventy to one hundred years, allowing LISA astronomers to watch them for years. Each object would serve as an exquisite probe for mapping the space-time geometry, the gravitational twists and turns, around a supermassive black hole.

All the while LISA will be listening to that cacophony of gravity-wave signals arriving from the host of binary star systems in our own Milky Way galaxy: the waves continually emitted as neutron stars orbit neutron stars, black holes circle black holes, white dwarfs pair up with other white dwarfs, and all the possible combinations in between. Many of these systems cannot be seen with regular telescopes, so gravity-wave detectors will offer the means of taking a reliable census of these binaries. Astronomers will hit the jackpot if, in one of those systems, a white dwarf merges with another white dwarf, generating a spectacular supernova. By one estimate, LISA has a 2 percent chance of seeing such an event over its lifetime.

But most thrilling of all will be the waves arriving from the primeval cosmos. As noted earlier, it is a space-based interferometer that will offer astronomers a particularly good opportunity to look directly back to our universe's origins, farther than with any other astronomical means. Here will be found the cosmic gravitational wave background or a view of the forces of nature emerging from scratch. "I think that in a couple of decades or so," said Kip Thorne, "it's likely that the most exciting things going on in gravitational-wave astronomy will be exploring what happened in the first second of the life of the universe."

The Gravitational Wave Spectrum

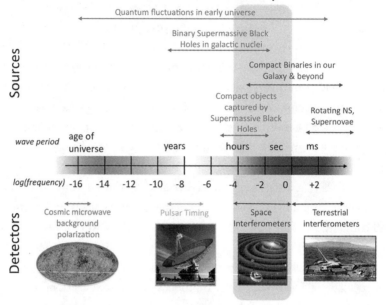

Much as astronomers now observe the universe across a wide range of electromagnetic frequencies—radio, microwaves, infrared, optical, ultraviolet, X-rays, and gamma rays—gravitational-wave physicists are either using or planning a variety of instruments and techniques, on Earth and in space, that will allow them to observe a broad band of gravitational-wave frequencies. This will enable them to study different types of celestial events. *NASA Goddard Space Flight Center.*

It was surprising and seemingly minor problems that caused intermittent delays in LIGO's startup in the late 1990s. At Hanford a few weeks had to be spent replacing a glue that failed, the glue that attached tiny maneuvering magnets to the first-generation mirrors. While the magnets were being reglued at Hanford, Weiss came out for a visit to measure any residues remaining after the arms had undergone their bake-out, which cleared the steel tubes of their last remaining gases. An electric current had been sent through the beam tubes as if they were long wires. For about a month the arms were heated in this way to 300° Fahrenheit. The total electric bill was more than sixty thousand dollars. Weiss set up his temporary office in a small portable trailer, parked right outside Port 5, an access door about halfway down the northern arm. He opened a series of valves to link his gas detector to the tube's interior and then patiently sat at a computer in the cramped quarters, keenly watching the screen as a graph gradually displayed the level of residual gases left in the tube. The first results were encouraging. "Isn't that pretty," he said, as the program drew its lines. "I don't see a lot from the tube." It was too good to be true, though. A half hour later he noticed a subtle change in a curve on his computer screen. It was the distinct signature of a leak. Thankfully, a test the next day confirmed that the leak was occurring at a port rather than at a failed weld, which would have been far more difficult to repair. A port leak can easily be fixed by tightening a bolt or replacing a gasket.

Weiss got a bit nostalgic sitting in the trailer. At that moment many years had passed since he first sat down and conceived of LIGO's basic design. Now the instrument stretched out before him, as miles of steel and concrete. "I was here when the first beam tubes were being installed," he recalled. "The meadowlarks and magpies would gather right outside. We'd also see swallows fly straight down the tubes, riding on the thermals." After a pause he continued. "A lot of heartache went into this, but it was all worthwhile."

Weiss, today in his eighties and ever the experimentalist, continues to travel to the LIGO observatories, rolling up his sleeves and checking on the equipment. "Every day on LIGO, there's a wonderful problem to work on," he says. "I sometimes look back at the things I've done—atomic clocks, laser stabilization, the cosmic microwave background—and all of them had one ingredient that was absolutely critical. They were fun to do." He never despaired, even after more than four long and turbulent decades helping bring LIGO to life: "The reason you don't worry about the end result is this—the problems were interesting, and you enjoyed the people you were working with. That's what keeps you going. If it isn't fun, then you give up. But it never came to that."

Weiss's current aspiration is to make sure that LIGO scientists can continue to place their ears upon the fabric of space-time and listen to the hidden rhythms of the universe. A few notes have registered—an opening tune. But in time, as sensitivities increase, there will be continual melodies, which eventually meld into a lush, resounding song. For gravity-wave astronomers today and into the future, Einstein's symphony will never be finished.

Coda

This day and age we're living in
Give cause for apprehension
With speed and new invention
And things like fourth dimension
Yet we get a trifle weary
With Mr. Einstein's theory
So we must get down to earth at times
Relax, relieve the tension
And no matter what the progress
Or what may yet be proved
The simple facts of life are such
They cannot be removed
You must remember this
A kiss is just a kiss
A sigh is just a sigh
The fundamental things apply
As time goes by . . .
—Original introduction to *As Time Goes By,* composed by Herman Hupfeld
for the 1931 Broadway show *Everybody's Welcome* and later used in the movie
Casablanca (Copyright © 1931 by Warner Brothers, Inc.)

TIMELINE

<table>
<tr><td>1687</td><td>Sir Isaac Newton publishes his revolutionary law of gravity in the Principia.</td></tr>
</table>

1687
Sir Isaac Newton publishes his revolutionary law of gravity in the *Principia*.

1816
The German mathematician Carl Friedrich Gauss discovers what he calls "non-Euclidean" geometries but doesn't officially publish his finding, although the insight makes him wonder whether physical space could be curved.

1829–32
The Russian mathematician Nikolai Lobachevsky and the Austro-Hungarian Janos Bolyai independently postulate a space of negative curvature, where more than one line through a point can be parallel to a given line.

1854
Gauss's student Bernhard Riemann introduces a geometry in which space has a positive curvature, where no parallel lines can be drawn through a point near a given line. He then generalizes the geometry of curved spaces into higher dimensions.

1905
Albert Einstein publishes what came to be known as his special theory of relativity, which abolishes Newtonian notions of absolute space and time.

1907
The Polish-German mathematician Hermann Minkowski demonstrates that Einstein's special relativity turned time into just another dimension, leading to the single entity of space-time.

1915
Introducing his general theory of relativity, Einstein successfully broadens relativity to handle other types of motion, specifically gravity. Gravity is now seen as arising from masses indenting the flexible mat of space-time, with objects moving along the curvatures. His equations were expressed in the language of Riemann's geometry.

1916
Einstein first discusses the concept of gravitational radiation, undulating waves in space-time generated as masses move about. He

publishes a follow-up paper on this idea two years later. Physicists begin arguing on whether such waves are real.

1919 British solar eclipse expeditions to West Africa and Brazil verify that starlight indeed bends its path as it passes close to the Sun, following the indentation the Sun makes in space-time. General relativity is triumphant.

1930 Subramanyan Chandrasekhar, then a graduate student at Cambridge University, calculates the maximum mass of a white-dwarf star, beyond which it must collapse.

1933 At a meeting of the American Physical Society, Caltech physicist Fritz Zwicky and Walter Baade of the Mount Wilson Observatory suggest that a tiny neutron star forms in the midst of a stellar explosion, a supernova.

1939 J. Robert Oppenheimer and his student George Volkoff at Berkeley are the first to examine the physics of a neutron star and discover that it can be stable. Afterward Oppenheimer joins with another student, Hartland Snyder, to publish the first modern description of a black hole. They call it "continued gravitational contraction."

1956 British theorist Felix Pirani publishes a paper in which he briefly mentions how, in a thought experiment, light signals could be used to detect gravitational waves. The debate on whether gravitational waves truly existed or were just an artifact of the mathematics is finally settled at a 1957 conference on gravitation at the University of North Carolina in Chapel Hill, where Pirani gave a talk on his ideas.

1960s University of Maryland physicist Joseph Weber becomes the first physicist to actively seek a gravity wave. He builds a series of resonant bar detectors, solid cylinders of metal surrounded by sensors to detect a gravity-wave-induced vibration.

1962 Two Russians, Mikhail E. Gertsenshtein and V. I. Pustovoit, first suggest using an interferometer instead of a resonant bar to detect gravitational waves. Others later come up with the idea independently, including Joseph Weber and MIT physicist Rainer Weiss.

1967 British astronomer Jocelyn Bell discovers pulsars, later understood to be spinning neutron stars only a dozen or so miles wide.

1969 Joseph Weber announces at a relativity conference that he has detected gravitational waves, possibly emanating from a neutron star. The announcement spurs physics groups around the world to build their own bars and mount similar searches.

1972 The seed for LIGO is planted. Rainer Weiss publishes the first serious study describing a kilometer-scale laser interferometer to detect gravitational waves and estimates all the possible sources of noise.

Robert Forward operates the first laser interferometer prototype, with two-meter-long arms, at Hughes Research Labs in Malibu, California.

1974   Radio astronomers Joseph Taylor and Russell Hulse, then at the University of Massachusetts, Amherst, find the first binary neutron-star system, labeled PSR 1913+16. The two neutron stars orbiting each other offer a perfect laboratory for tests on general relativity.

Inspired by the work of Rainer Weiss and Robert Forward, German physicists begin work on their own laser interferometer prototypes under the guidance of Heinz Billing. Into the 1980s they hold all the records for sensitivity for such an instrument, inventing techniques now standard in the field.

Joseph Weber's claims to have discovered gravitational waves are roundly contradicted at the Seventh International Conference on General Relativity and Gravitation held in Tel Aviv.

1978   After years of monitoring the binary pulsar PSR 1913+16, Joseph Taylor and several colleagues detect that its orbit is shrinking due to energy losses. The measured loss matched exactly what general relativity predicted if the system were losing energy in the form of gravity waves alone. It was the first indirect evidence that gravitational waves exist, leading to a Nobel prize for Taylor and Russell Hulse in 1993.

1979   The National Science Foundation funds both a gravitational-wave detection group at Caltech and a 40-meter laser interferometer on campus. Scottish physicist Ronald Drever is put in charge.

1982   The German physicist Roland Schilling conceives of power recycling, which helps laser interferometry become a serious contender as a gravitational-wave detection method.

1983   With Peter Saulson and Paul Linsay, Rainer Weiss completes a feasibility study for a kilometer-scale interferometer, which convinces the NSF to initiate research.

Ronald Drever coinvents a means to keep lasers stable, a necessity in pushing laser interferometry forward.

1984   LIGO is officially founded as a joint Caltech/MIT venture. The troika—Rainer Weiss at MIT, Kip Thorne and Ronald Drever at Caltech—oversee the new collaboration for the first three years.

Early 1990s   The National Science Board approves the LIGO construction proposal, which envisions two initial interferometers that would eventually be upgraded. Congress agrees to the funding.

1992   The NSF selects the LIGO detectors to be placed in Hanford, Washington, and Livingston, Louisiana.

1995   British and German physicists collaborate to build a six-hundred-meter laser interferometer south of Hannover, Germany. Known as GEO600, it primarily operates as a test-bed for technological innovations in laser interferometry.

1996   Italy and France begin construction of a LIGO-like gravitational wave detector known as Virgo, situated near Pisa.

1999	Construction of the initial LIGO detectors is completed and inauguration ceremonies are held.
2001	Initial LIGO begins to fully operate.
2010	Initial LIGO completes its mission, finding no gravitational waves over those years. Installation of Advanced LIGO equipment begins, which aims for an eventual tenfold increase in LIGO's sensitivity.
2014	The installation of Advanced LIGO is completed. A series of engineering tests begin.
Sept. 14, 2015	The first direct detection of gravitational waves. During an engineering run, days before the first observing search was to begin, Advanced LIGO records strong gravitational waves from the collision of two black holes located 1.3 billion light-years away. The event is labeled GW150914.
Dec. 26, 2015	LIGO makes its second bona fide detection of gravitational waves, also from two black holes colliding.
Feb. 11, 2016	LIGO's discovery of GW150914 is officially announced and published in *Physical Review Letters*.

BIBLIOGRAPHY

Aasi, J., et al. "Gravitational Waves from Known Pulsars: Results from the Initial Detector Era." *Astrophysical Journal* 785 (April 20, 2014): 119–136.

Abbott, B. P., et al. "Observation of Gravitational Waves from a Binary Black Hole Merger." *Physical Review Letters* 116, 061102 (2016).

———. "Properties of the Binary Black Hole Merger GW150914." *Physical Review Letters* 116, 241102 (2016).

Abbott, David, ed. *Mathematicians.* London: Blond Educational, 1985.

Aglietta, M., et al. "Correlation between the Maryland and Rome Gravitational-Wave Detectors and the Mont Blanc, Kamioka and IMB Particle Detectors during SN 1987A." *Il Nuovo Cimento* 106 (November 1991): 1257–1269.

Allen, B., et al. "White Paper Outlining the Data Analysis System (DAS) for LIGO I." Document LIGO-M970065-B, June 9, 1997.

Anderson, Christopher. "Divorce Splits LIGO's 'Dysfunctional Family.'" *Science* 260 (May 21, 1993): 1063.

———. "LIGO Director Out in Shakeup." *Science* 263 (March 11, 1994): 1366.

Angel, Roger B. *Relativity: The Theory and Its Philosophy.* Oxford: Pergamon, 1980.

Baker, John G., et al. "Gravitational-Wave Extraction from an Inspiraling Configuration of Merging Black Holes." *Physical Review Letters* 96, 111102 (March 24, 2006): 1–4.

Barish, Barry, and Rainer Weiss. "Gravitational Waves Really Are Shifty." *Physics Today* 53 (March 2000): 105.

———. "LIGO and the Detection of Gravitational Waves." *Physics Today* 52 (October 1999): 44–50.

Bartusiak, Marcia. "Celestial Zoo." *Omni* 5 (December 1982): 106–113.

———. "Einstein's Unfinished Symphony." *Discover* 10 (August 1989): 62–69.

———. "Experimental Relativity: Its Day in the Sun." *Science News* 116 (August 25, 1979): 140–142.

——. "Gravity Wave Sky." *Discover* 14 (July 1993): 72–77.

——. "Sensing the Ripples in Space-Time." *Science 85* 6 (April 1985): 58–65.

Begelman, Mitchell, and Martin Rees. *Gravity's Fatal Attraction*. New York: Scientific American Library, 1996.

Bell, E. T. *Men of Mathematics*. New York: Simon and Schuster, 1965.

Bender, P., et al. (LISA Study Team.) *LISA: Laser Interferometer Space Antenna for the Detection and Observation of Gravitational Waves*. Pre-Phase A Report, 2nd ed. Garching, Germany: Max-Planck-Institut für Quantenoptik, July 1998.

Blair, David, and Geoff McNamara. *Ripples on a Cosmic Sea*. Reading, Mass.: Helix Books, Addison-Wesley, 1997.

Bromberg, Joan Lisa. *The Laser in America, 1950–1970*. Cambridge, Mass.: MIT Press, 1991.

Burnell, Jocelyn Bell. "The Discovery of Pulsars." In *Serendipitous Discoveries in Radio Astronomy*. Proceedings of Workshop Number 7 held at the National Radio Astronomy Observatory, Green Bank, W.Va., May 4–6, 1983.

Campanelli, M., C. O. Lousto, P. Marronetti, and Y. Zlochower. "Accurate Evolutions of Orbiting Black-Hole Binaries with Excision." *Physical Review Letters* 96, 111101 (March 24, 2006): 1–4.

Carmeli, Moshe, Stuart I. Fickler, and Louis Witten, eds. *Relativity: Proceedings of the Relativity Conference in the Midwest*. New York: Plenum, 1970.

Chandrasekhar, S. *Eddington*. Cambridge: Cambridge University Press, 1983.

Ciufolini, Ignazio, and John Archibald Wheeler. *Gravitation and Inertia*. Princeton, N.J.: Princeton University Press, 1995.

Cohen, I. Bernard. *Roemer and the First Determination of the Velocity of Light*. New York: Burndy Library, 1944.

Coles, Mark. "How the LIGO Livingston Observatory Got Its Name." *Latest from LIGO Newsletter* 3 (April 1998).

Collins, H. M. *Changing Order: Replication and Induction in Scientific Practice*. Chicago: University of Chicago Press, 1992.

——. "A Strong Confirmation of the Experimenters' Regress." *Studies in History and Philosophy of Science* 25 (1994): 493–503.

Committee on Gravitational Physics, National Research Council. *Gravitational Physics: Exploring the Structure of Space and Time*. Washington, D.C.: National Academy Press, 1999.

Connaughton, V., et al. "*Fermi* GBM Observations of LIGO Gravitational-Wave Event GW150914." *Astrophysical Journal Letters* 826 (July 20, 2016): L6–L33.

Coyne, Dennis. "The Laser Interferometer Gravitational-Wave Observatory (LIGO) Project." In *Proceedings of the IEEE 1996 Aerospace Applications Conference* 4 (1996): 31–61.

Cutler, Curt, and Kip S. Thorne. "An Overview of Gravitational-Wave Sources." In *Proceedings of the 16th International Conference on General Relativity and Gravitation*, ed. Nigel T. Bishop and Sunil D. Maharaj. Singapore: World Scientific, 2002.

Cyranoski, David. "Chinese Gravitational-Wave Hunt Hits Crunch Time." *Nature* 531 (March 10, 2016): 150–51.

D'Abro, A. *The Evolution of Scientific Thought: From Newton to Einstein*. New York: Dover, 1950.

Damour, T. "Theoretical Aspects of Gravitational Radiation." In *General Relativity and Gravitation: Proceedings of the 14th International Conference*. Singapore: World Scientific, 1997.

Davies, P. C. W. *The Search for Gravity Waves*. Cambridge: Cambridge University Press, 1980.

——. *Space and Time in the Modern Universe*. Cambridge: Cambridge University Press, 1977.

Dietrich, Jane. "Realizing LIGO." *Engineering and Science* 61, no. 2 (1998): 8–17.

Dreyer, R. W. P. "Gravitational Wave Astronomy." *Quarterly Journal of the Royal Astronomical Society* 18 (1977): 9–27.

Dunnington, G. Waldo. *Carl Friedrich Gauss: Titan of Science*. New York: Exposition Press, 1955.

Dyson, Freeman. "Gravitational Machines." In *Interstellar Communication*. New York: W. A. Benjamin, 1963.

Eddington, A. S. *The Mathematical Theory of Relativity*. Cambridge: Cambridge University Press, 1930.

——. *Space, Time and Gravitation*. Cambridge: Cambridge University Press, 1920.

——. *The Theory of Relativity and Its Influence on Scientific Thought* (The Romanes Lecture, 1922). Oxford: Clarendon Press, 1922.

Einstein, Albert. "Die Grundlagen der allgemeinen Relativitätstheorie." *Annalen der Physik* 49 (1916): 769–822.

——. *Relativity: The Special and the General Theory*. New York: Crown Trade, 1961.

——. *Sitzungsberichte der Königlich Preussichen Akademie der Wissenschaften* (1916): 688.

Ellis, George F. R., and Ruth M. Williams. *Flat and Curved Space-Times*. Oxford: Clarendon Press, 1988.

Ezrow, D. H., N. S. Wall, J. Weber, and G. B. Yodh. "Insensitivity to Cosmic Rays of the Gravity Radiation Detector." *Physical Review Letters* 24 (April 27, 1970): 945–947.

Faber, Scott. "Gravity's Secret Signals." *New Scientist* 144 (November 26, 1994): 40.

Fauvel, John, Raymond Flood, Michael Shortland, and Robin Wilson, eds. *Let Newton Be!* Oxford: Oxford University Press, 1988.

Fermi, Laura, and Gilberto Bernardini. *Galileo and the Scientific Revolution*. New York: Basic Books, 1961.

Ferrari, V., G. Pizzella, M. Lee, and J. Weber. "Search for Correlations between the University of Maryland and the University of Rome Gravitational Radiation Antennas." *Physical Review D* 25 (May 15, 1982): 2471–2486.

Finn, Lee Samuel. "A Numerical Approach to Binary Black Hole Coalescence." In *General Relativity and Gravitation: Proceedings of the 14th International Conference*. Singapore: World Scientific, 1997.

Fisher, Arthur. "The Tantalizing Quest for Gravity Waves." *Popular Science* 218 (April 1980): 88–94.

Flam, Faye. "A Prize for Patient Listening." *Science* 262 (October 22, 1993): 507.

——. "Scientists Chase Gravity's Rainbow." *Science* 260 (April 23, 1993): 493.

Florence, Ronald. *The Perfect Machine: Building the Palomar Telescope.* New York: HarperCollins, 1994.

Folkner, William M., ed. *Laser Interferometer Space Antenna: Second International LISA Symposium on the Detection and Observation of Gravitational Waves in Space.* AIP Conference Proceedings 456. Woodbury, N.Y.: American Institute of Physics, 1998.

Fölsing, Albrecht. *Albert Einstein: A Biography.* New York: Viking, 1997.

Fomalont, E. B., and R. A. Sramek. "Measurements of the Solar Gravitational Deflection of Radio Waves in Agreement with General Relativity." *Physical Review Letters* 36 (June 21, 1976): 1475–1478.

Forward, Robert L. "Wideband Laser-Interferometer Experiment." *Physical Review D* 17 (January 15, 1978): 379–390.

Franklin, Allan. "How to Avoid the Experimenters' Regress." *Studies in History and Philosophy of Science* 25 (1994): 463–491.

Friedman, Michael. *Foundations of Space-Time Theories: Relativistic Physics and Philosophy of Science.* Princeton, N.J.: Princeton University Press, 1983.

Garisto, Robert. "How Gravitational Waves Went from a Whisper to a Shout." *Physics Today* 69 (August 2016): 10–11.

Garwin, Richard L. "Detection of Gravity Waves Challenged." *Physics Today* 27 (December 1974): 9–11.

——. "More on Gravity Waves." *Physics Today* 28 (November 1975): 13.

Garwin, Richard L., and James L. Levine. "Single Gravity-Wave Detector Results Contrasted with Previous Coincidence Detections." *Physical Review Letters* 31 (July 16, 1973): 176–180.

Gertsenshtein, M. E., and V. I. Pustovoit. "On the Detection of Low Frequency Gravitational Waves." *Soviet Physics JETP* 16 (February 1963): 433–435.

Giazotto, Adalberto. "Interferometric Detection of Gravitational Waves." *Physics Reports* 182 (November 1989): 365–424.

Gibbons, G. W., and S. W. Hawking. "Theory of the Detection of Short Bursts of Gravitational Radiation." *Physical Review D* 4 (October 15, 1971): 2191–2197.

Gillispie, Charles Coulston, ed. *Dictionary of Scientific Biography.* New York: Charles Scribner's Sons, 1972.

Glanz, James. "Gamma Blast from Way, Way Back." *Science* 280 (April 24, 1998): 514.

Gorman, Peter. *Pythagoras: A Life.* London: Routledge and Kegan Paul, 1979.

"Gravitating toward Einstein." *Time* 93 (June 20, 1969): 75.

"Gravitational Waves Detected." *Science News* 95 (June 21, 1969): 593–594.

"Gravitational Waves Detected." *Sky and Telescope* 38 (August 1969): 71.

"Gravity Waves Slow Binary Pulsar." *Physics Today* 32 (May 1979): 19–20.

Hall, Tord. *Carl Friedrich Gauss.* Cambridge, Mass.: MIT Press, 1970.

Hariharan, P. *Optical Interferometry.* Sydney: Academic Press, 1985.

Hawking, S. W., and W. Israel, eds. *General Relativity: An Einstein Centenary Survey.* Cambridge: Cambridge University Press, 1979.

Hilts, Philip J. "Last Rites for a 'Plywood Palace' That Was a Rock of Science." *New York Times*, March 31, 1998, C4.

Hinckfuss, Ian. *The Existence of Space and Time*. Oxford: Clarendon Press, 1975.

Hoffmann, Banesh. *Albert Einstein: Creator and Rebel*. New York: Viking, 1972.

——. *Relativity and Its Roots*. New York: W. H. Freeman, 1983.

Hulse, Russell A. "The Discovery of the Binary Pulsar." *Reviews of Modern Physics* 66 (July 1994): 699–710.

Jammer, Max. *Concepts of Space*. Cambridge, Mass.: Harvard University Press, 1954.

Kepler, Johannes. *The Harmony of the World*. Philadelphia: American Philosophical Society, 1997.

Kimball, Robert, ed. *The Complete Lyrics of Cole Porter*. New York: Alfred A. Knopf, 1983.

Kleppner, Daniel. "The Gem of General Relativity." *Physics Today* 46 (April 1993): 9–11.

Laser Pioneer Interviews. Torrance, Calif.: High Tech Publications, 1985.

Lee, M., D. Gretz, S. Steppel, and J. Weber. "Gravitational-Radiation-Detector Observations in 1973 and 1974." *Physical Review D* 14 (August 15, 1976): 893–906.

Levine, James L., and Richard L. Garwin. "Absence of Gravity-Wave Signals in a Bar at 1695 Hz." *Physical Review Letters* 31 (July 16, 1973): 173–176.

LIGO Magazine 8 (March 2016): http://www.ligo.org/magazine/LIGO-magazine-issue-8.pdf.

Lubkin, Gloria B. "Experimental Relativity Hits the Big Time." *Physics Today* 23(August 1970): 41–44.

——. "Weber Reports 1660-Hz Gravitational Waves from Outer Space." *Physics Today* 22 (August 1969): 61–62.

Machamer, Peter K., and Robert G. Turnbull, eds. *Motion and Time/Space and Matter: Interrelations in the History of Philosophy and Science*. Columbus: Ohio State University Press, 1976.

Marshall, Eliot. "Garwin and Weber's Waves." *Science* 212 (May 15, 1981): 765.

Mather, John C., and John Boslough. *The Very First Light*. New York: Basic Books, 1996.

Michelson, Peter F., John C. Price, and Robert C. Taber. "Resonant-Mass Detectors of Gravitational Radiation." *Science* 237 (July 10, 1987): 150–157.

Minkowski, H. "Das Relativitätsprinzip." *Jahresbericht der Deutschen Mathematiker-Vereinigung* 24 (1915): 372–382.

Misner, C. W., K. S. Thorne, and J. A. Wheeler. *Gravitation*. San Francisco: W. H. Freeman, 1973.

Monastyrsky, Michael. *Riemann, Topology, and Physics*. Boston: Birkhauser, 1987.

Moskowitz, Eric. "The Chirp That Proved Einstein Right." *Boston Globe*, May 15, 2016, A1, A8–A9.

Moss, G. E., L. R. Miller, and R. L. Forward. "Photon-Noise-Limited Laser Transducer for Gravitational Antenna." *Applied Optics* 10 (November 1971): 2495–2498.

Naeye, Robert. "To Catch a Gravity Wave." *Discover* 14 (July 1993): 76.

Norton, John. "How Einstein Found His Field Equations, 1912–1915." In *Einstein and the History of General Relativity*, ed. Don Howard and John Stachel. Boston: Birkhauser, 1989.

Nuttall, Laura, and Jess McIver. "Understanding the Characteristics of the LIGO Detectors." *LIGO Magazine* 8 (March 2016): 24–25.

Overbye, Dennis. "Testing Gravity in a Binary Pulsar." *Sky and Telescope* 57 (February 1979): 146.

Pais, Abraham. *"Subtle Is the Lord . . .": The Science and the Life of Albert Einstein.* Oxford: Oxford University Press, 1982.

Park, David. *The How and the Why.* Princeton, N.J.: Princeton University Press, 1988.

Perkowitz, Sidney. *Empire of Light.* New York: Henry Holt, 1996.

Perryman, Michael. "Hipparcos: The Stars in Three Dimensions." *Sky and Telescope* (June 1999): 40–50.

Pirani, E. A. E. "On the Physical Significance of the Riemann Tensor." *Acta Physica Polonica* 15 (1956): 389–405.

Pound, R. V., and G. A. Rebka Jr. "Apparent Weight of Photons." *Physical Review Letters* 4 (April 1, 1960): 337–340.

Preparata, Giuliano. "Joe Weber's Physics." In *QED Coherence in Matter.* Singapore: World Scientific, 1995.

Press, William H., and Kip S. Thorne. "Gravitational-Wave Astronomy." *Annual Review of Astronomy and Astrophysics* 10 (1972): 335–374.

Preston, Richard. *First Light: The Search for the Edge of the Universe.* New York: Atlantic Monthly, 1987.

Pretorius, Frans. "Evolution of Binary Black-Hole Spacetimes." *Physical Review Letters* 95, 121101 (September 16, 2005): 1–4.

———. "Numerical Relativity Using a Generalized Harmonic Decomposition." *Classical and Quantum Gravity* 22 (2005): 425–51.

Raine, Derek J., and Michael Heller. *The Science of Space-Time.* Tucson, Ariz.: Pachart, 1981.

Salmon, Wesley C. *Space. Time, and Motion.* Minneapolis: University of Minnesota Press, 1980.

Sanders, J. H. *The Velocity of Light.* Oxford: Pergamon, 1965.

Saulson, Peter R. *Fundamentals of Interferometric Gravitational Wave Detectors.* Singapore: World Scientific, 1994.

———. "If Light Waves Are Stretched by Gravitational Waves, How Can We Use Light as a Ruler to Detect Gravitational Waves?" *American Journal of Physics* 65 (June 1997): 501–505.

———. "Josh Goldberg and the Physical Reality of Gravitational Wave." *General Relativity and Gravitation* 43 (December 2011): 3289–3299.

Schilpp, Paul A., ed. *Albert Einstein: Philosopher-Scientist.* New York: Harper Torchbooks, 1959.

Schwinger, Julian. *Einstein's Legacy.* New York: Scientific American Books, 1986.

Shapiro, Irwin I. "A Century of Relativity." In *More Things in Heaven and Earth: A Celebration of Physics at the Millennium,* ed. Benjamin Bederson. New York: Springer, 1999.

Shapiro, Irwin I. "Fourth Test of General Relativity." *Physical Review Letters* 13 (December 28, 1964): 789–791.

Shapiro, Irwin I., et al. "Fourth Test of General Relativity: New Radar Result." *Physical Review Letters* 26 (May 3, 1971): 1132–1135.

———. "Fourth Test of General Relativity: Preliminary Results." *Physical Review Letters* 20 (May 27, 1968): 1265–1269.

Shapiro, Stuart L., Richard F. Stark, and Saul A. Teukolsky. "The Search for Gravitational Waves." *American Scientist* 73 (May–June 1985): 248–257.

Shaviv, G., and J. Rosen, eds. *General Relativity and Gravitation: Proceedings of the Seventh International Conference (GR7)*. New York: John Wiley and Sons, 1975.

Silk, Joseph. "Black Holes and Time Warps: Einstein's Outrageous Legacy." *Science* 264 (May 13, 1994): 999.

Sklar, Lawrence. *Space, Time, and Spacetime*. Berkeley: University of California Press, 1974.

Smart, J. J. C., ed. *Problems of Space and Time*. New York: Macmillan, 1964.

Sobel, Michael I. *Light*. Chicago: University of Chicago Press, 1987.

Stein, Mark A. "Leggo My LIGO: Feud between Two Key Researchers Rattles $250-Million Gravity-Wave Project; Officials Say It Is Still on Track." *Los Angeles Times*, April 28, 1993, B1.

Taubes, Gary. "The Gravity Probe." *Discover* 18 (March 1997): 62.

Taylor, Edwin F., and John Archibald Wheeler. *Spacetime Physics*. New York: W. H. Freeman, 1966.

Taylor, Joseph H., Jr. "Binary Pulsars and Relativistic Gravity." *Reviews of Modern Physics* 66 (July 1994): 711–719.

Thomsen, Dietrick. "Gaining on Gravity Waves." *Science News* 98 (July 11, 1970): 44–45.

———. "Trying to Rock with Gravity's Vibes." *Science News* 126 (August 4, 1984): 76–78.

Thorne, Kip. *Black Holes and Time Warps: Einstein's Outrageous Legacy*. New York: W. W. Norton, 1994.

———. "Gravitational-Wave Research: Current Status and Future Prospects." *Reviews of Modern Physics* 52 (April 1980):285–297.

Torretti, Roberto. *Relativity and Geometry*. Oxford: Pergamon, 1983.

Tourrene, Philippe. *Relativity and Gravitation*. Cambridge: Cambridge University Press, 1997.

Travis, John. "LIGO: A $250 Million Gamble." *Science* 260 (April 30, 1993): 612–614.

Trimble, Virginia, and Joseph Weber. "Gravitational Radiation Detection Experiments with Disc-Shaped and Cylindrical Antennae and the Lunar Surface Gravimeter." *Annals of the New York Academy of Sciences* 224 (December 14, 1973): 93–107.

Tyson, J. Anthony. "Gravitational Radiation." *Annals of the New York Academy of Sciences* 224 (December 14, 1973): 74–92.

———. Testimony before the Subcommittee on Science of the Committee on Science, Space, and Technology, US House of Representatives, March 13, 1991.

Vessot, R. F. C., and M. W. Levine. "A Test of the Equivalence Principle Using a Space-Borne Clock." *General Relativity and Gravitation* 10 (1979): 181–201.

Vessot, R. F. C., et al. "Test of Relativistic Gravitation with a Space-Borne Hydrogen Maser." *Physical Review Letters* 45 (December 29, 1980): 2081–2084.

Wald, Robert M. *Space, Time, and Gravity.* Chicago: University of Chicago Press, 1992.

Waldrop, M. Mitchell. "Of Politics, Pulsars, Death Spirals—and LIGO." *Science* 249 (September 7, 1990): 1106.

Weber, Joseph. "Anisotropy and Polarization in the Gravitational-Radiation Experiments." *Physical Review Letters* 25 (July 20, 1970): 180–184.

——. "Detection and Generation of Gravitational Waves." *Physical Review* 117 (January 1, 1960): 306–313.

——. 'The Detection of Gravitational Waves." *Scientific American* 224 (May 1971): 22–29.

——. "Evidence for Discovery of Gravitational Radiation." *Physical Review Letters* 22 (June 16, 1969): 1320–1324.

——. "Gravitational Radiation Detector Observations in 1973 and 1974." *Nature* 266 (March 17, 1977): 243.

——. "Gravitational Waves." *Physics Today* 21 (April 1968): 34–39.

——. "Gravitons, Neutrinos, and Antineutrinos." *Foundations of Physics* 14 (December 1984): 1185–1209.

——. "How I Discovered Gravitational Waves." *Popular Science* 200 (May 1972): 106+.

——. "Weber Replies." *Physics Today* 27 (December 1974): 11–13.

——. "Weber Responds." *Physics Today* 28 (November 1975): 13+.

Weber, J., and T. M. Karade, eds. *Gravitational Radiation and Relativity.* Singapore: World Scientific, 1986.

Weber, J., and B. Radak. "Search for Correlations of Gamma-Ray Bursts with Gravitational-Radiation Antenna Pulses." *Il Nuovo Cimento* 111B (June 1996): 687–692.

Weisberg, Joel M., and Joseph H. Taylor. "The Relativistic Binary Pulsar B1913+16: Thirty Years of Observations and Analysis." In *Binary Radio Pulsars*, ASP Conference Series, ed. F. A. Rasio and I. H. Stairs, vol. 328 (2005), 25.

Weiss, Rainer. "Gravitational Radiation." In *More Things in Heaven and Earth: A Celebration of Physics at the Millennium*, ed. Benjamin Bederson. New York: Springer, 1999.

——. "Gravitation Research." Massachusetts Institute of Technology. Research Laboratory of Electronics. *Quarterly Progress Report* 105 (1972): 54–76.

Westfall, Richard S. *The Life of Isaac Newton.* Cambridge: Cambridge University Press, 1993.

Wheeler, John Archibald, and Kenneth Ford. *Geons, Black Holes, and Quantum Foam.* New York: W. W. Norton, 1998.

Will, Clifford M. "The Confrontation Between Gravitation Theory and Experiment." In *General Relativity: An Einstein Centenary Survey.* Cambridge: Cambridge University Press, 1979.

———. "General Relativity at 75: How Right Was Einstein?" *Science* 250 (November 9, 1990): 770–776.

———. "Gravitational Radiation and the Validity of General Relativity." *Physics Today* 52 (October 1999): 38–43.

———. *Was Einstein Right?* New York: Basic Books, 1986.

Williams, L. Pearce, ed. *Relativity Theory: Its Origins and Impact on Modern Thought.* Huntington, N.Y.: Robert E. Krieger, 1979.

INDEX

Note: Page numbers in italics refer to illustrations.